Project AIR FORCE

# Supporting Expeditionary Aerospace Forces

# AN ANALYSIS OF F-15 AVIONICS OPTIONS

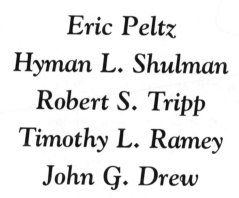

*Eric Peltz*

*Hyman L. Shulman*

*Robert S. Tripp*

*Timothy L. Ramey*

*John G. Drew*

Prepared for the
UNITED STATES AIR FORCE

## RAND

The research reported here was sponsored by the United States Air Force under Contract F49642-96-C-0001. Further information may be obtained from the Strategic Planning Division, Directorate of Plans, Hq USAF.

**Library of Congress Cataloging-in-Publication Data**

Supporting expeditionary aerospace forces : an analysis of F-15 avionics options / Eric Peltz ... [et al,].
    p. cm.
    "MR-1174-AF."
    Includes bibliographical references.
    ISBN 0-8330-2905-3
    1. Avionics—United States. 2. Eagle (Jet fighter plane)  I. Peltz, Eric, 1968–

UG1423 .S87 2000
358.4'3'0973—dc21

00-062666

Published 2000 by RAND
Street, P.O. Box 2138, Santa Monica, CA 90407-2138
uth Hayes Street, Arlington, VA 22202-5050
'AND URL: http://www.rand.org/
cuments or to obtain additional information,
tion Services: Telephone: (310) 451-7002;
(310) 451-6915; E-mail: order@rand.org

This report discusses the manner in which emerging Air Force employment strategies for the Expeditionary Aerospace Force (EAF) should be supported. Although much work remains to define and prepare Air Force units for EAF responsibilities, it is clear that EAF concepts will play a critical role in the future Air Force. EAF concepts rely on rapidly deployable, immediately employable, highly effective, and flexible air and space force packages that can fill the same strategic role as a permanent forward presence in deterring and responding to aggression. To a great extent, EAF success will depend on the effectiveness and efficiency of the support system that undergirds combat operations. The Air Force has designated such a support system one of its six necessary core competencies and has labeled it the Agile Combat Support (ACS) system.

Planning, programming, and budgetary system decisions affect both the efficiency and the effectiveness of ACS systems. Long-term ACS decisions in turn influence the support structures needed to meet the operational requirements of future force mixes. Midterm ACS decisions affect the design, development, and evolution of the support infrastructure for meeting operational requirements within programming and budgeting time horizons. Near-term decisions affect where, when, and how existing resources are employed. At all such stages, logistics requirements can be satisfied in a variety of ways, each associated with different costs, degrees of flexibility, response times, and risks.

This study addresses alternatives for meeting demands for F-15 avionics components across the spectrum of EAF operations, ranging from major theater wars to peacetime operations. The options (strategic decisions) available range from the current policy of deploying intermediate-maintenance capabilities with the flying units to decentralized and consolidated efforts in which repairable avionics components are transported between units and maintenance locations.

Our research shows that consolidating intermediate F-15 avionics maintenance offers the potential to significantly enhance support responsiveness, reduce deployment airlift requirements, and ease the stress that frequent and unpredictable deployments place on maintenance personnel.

This research, sponsored by the Air Force Deputy Chief of Staff for Installations and Logistics (AF/IL), was conducted by the Resource Management Program of Project AIR FORCE. It is part of a larger project on "Evaluating Agile Combat Support Options for Implementing the Expeditionary Air Force." Other project publications include *Supporting Expeditionary Aerospace Forces: A Concept for Evolving the Agile Combat Support/Mobility System of the Future*, MR-1179-AF, 2000; *Supporting Expeditionary Aerospace Forces: An Integrated Strategic Agile Combat Support Planning Framework*, MR-1056-AF, 1999; *Supporting Expeditionary Aerospace Forces: Expanded Analysis of LANTIRN Options*, MR-1225-AF, 2000; and *Supporting Expeditionary Aerospace Forces: New Agile Combat Support Postures*, MR-1075-AF, 2000.

This report examines the logistics processes and organizational design for F-15 avionics maintenance and assumes the reader is familiar with Air Force logistics. Those less familiar with Air Force logistics or logistics in general should still find Chapter One (Introduction), Chapter Two (Support Structure Options and the Decision Space), and Chapter Seven (Conclusion) informative. For a more general treatment of the implications of this analysis and related works, we recommend MR-1179-AF, *Supporting Expeditionary Aerospace Forces: A Concept for Evolving the Agile Combat Support/Mobility System of the Future*.

## PROJECT AIR FORCE

Project AIR FORCE, a division of RAND, is the Air Force Federally Funded Research and Development Center (FFRDC) for studies and analyses. It provides the Air Force with independent analyses of policy alternatives affecting the development, employment, combat readiness, and support of current and future aerospace forces. Research is performed in four programs: Aerospace Force Development; Manpower, Personnel, and Training; Resource Management; and Strategy and Doctrine.

# CONTENTS

# FIGURES

# TABLES

In the current Air Force support system for F-15 avionics, each base with F-15 aircraft has an avionics intermediate maintenance shop (AIS) for repairing avionics line replaceable units (LRUs), or components that are removed and replaced by flight line mechanics. The present policy is to deploy the AIS with aircraft from home bases to forward operating locations (FOLs). We refer to this concept as a decentralized-deployment support option. This system places a heavy deployment burden on avionics personnel and requires substantial airlift for the AIS equipment. These burdens also adversely affect the Expeditionary Aerospace Force (EAF) goals of increasing response speed, reducing strain on personnel, and reducing deployment footprint, or the amount of materiel that must deploy with a force.

What F-15 avionics maintenance options might the Air Force consider in its efforts to achieve EAF goals? We examine alternatives that eliminate or reduce AIS deployments by providing spare parts replenishments to FOLs through distribution rather than local repair, comparing these alternatives both to each other and to the current system. These include:

- The current decentralized-deployment system

- A decentralized-no-deployment system in which each AIS supports deployed aircraft from home instead of deploying with aircraft to FOLs

- A single continental United States (CONUS) support location (CSL) with consolidated repair for worldwide support in both peace and war

- A CSL in network with two, three, or four regional repair forward support locations (FSL) that would support operations in both peace and war.

Figure S.1 presents a notional support structure comprising four FSLs and one CSL located at existing and hypothetical bases.

Our analysis focuses on system costs, deployment requirements, and operational risks associated with each of these these alternatives. We also consider how technological change and transformation of current processes would affect system performance in meeting EAF goals. We consider, for example, how faster order-and-ship times than those currently achieved and implementation of the electronic system test set (ESTS) being developed to reduce deployment footprint and personnel requirements would affect comparisons between support structure alternatives.

RAND *MR1174-AF-S.1*

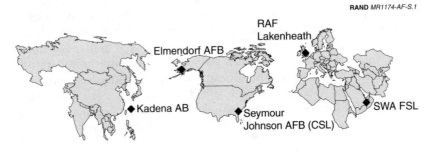

◆ Consolidated support location

**Figure S.1—Notional Four-FSL and One-CSL Structure**

## SYSTEM COSTS

To compare the costs of the alternatives, we calculated the present value of operating and investment costs. We found that the consolidated alternatives reduce annual operating costs in exchange for initial investments in F-15 avionics serviceable spare parts. The level of consolidation affects the balance of this tradeoff in that greater consolidation yields the greatest reduction in personnel costs, but this gain is offset by even greater increases in spare parts requirements, as well as by smaller increases in transportation costs. The net result is that, using the current testers and assuming current order-and-ship times, the four-FSL/one-CSL option yields the lowest net cost of the consolidated alternatives and is the only alternative that is cost-competitive with the current decentralized-deployment system. Reducing order-and-ship times would make the four-FSL/one-CSL system less costly than the current system.

Each alternative using the ESTS and current order-and-ship times would be more costly than the current system using the ESTS. This is because adoption of ESTS in itself produces some of the personnel savings generated by consolidation, the biggest cost advantage of consolidated systems. With the ESTS, reducing order-and-ship times makes the four-FSL/one-CSL option only slightly more expensive than the current system.

## REDUCING EQUIPMENT AND PERSONNEL DEPLOYMENT REQUIREMENTS

Quick-hitting expeditionary operations require rapid deployment of combat forces, placing a premium on reducing deployment footprint, or the amount of initial airlift needed to transport support equipment. For a major theater war deployment, all the alternatives we considered would reduce deployment footprint for F-15 avionics maintenance capabilities by up to 60 C-141 or 43 C-17 load equivalents from that needed for the current decentralized-deployment system. ESTS adoption would also greatly reduce deployment footprint for the current system, so with ESTS the alternatives we considered would generate only a marginal reduction in deployment footprint of 12 C-141 or 9 C-17 load equivalents.

The current decentralized structure has high and frequent personnel deployment requirements. The consolidated structures would eliminate deployment requirements for some small-scale contingencies and reduce them for major theater wars. Each consolidated alternative we consider would also be less stressing than those required by the current system in that deployments would be to FSLs rather than to FOLs, which are more likely to be in hostile areas. Of course, the decentralized-no-deployment structure would eliminate deployment personnel requirements.

Personnel retention problems have made it difficult for the Air Force to maintain the required skill-level mix of personnel in areas such as F-15 avionics repair. To solve this problem, the Air Force can either work toward improving the retention of its current personnel or find other sources of repair personnel. The Air Force has attributed its personnel retention problems to frequent deployments to FOLs over the last decade. RAND research on the effects of deployment on personnel retention conceptually supports this contention but also concludes that a low to moderate level of deployment, particularly to nonhostile locations such as those in which FSLs would be positioned, has a positive effect on personnel retention.[1] By this standard, a CSL in network with FSLs would be the most favorable alternative for personnel retention, but the elimination of deployments would probably remain preferable to excessive deployments to FOLs in potentially hostile environments. Alternatively, the elimination of deployments to FOLs gives the Air Force flexibility in how it decides to achieve required personnel levels. Should the Air Force seek to find other sources of repair personnel, eliminating deployments or keeping them limited to FSLs would allow for the use of contractors, government-employed civilians, or allied partnerships.

## RISK

Decentralized and consolidated structures carry different operational risks. Decentralized deployment is associated with risks in equipment deployment, setup, and downtime. Current planning assumes that the AIS will deploy and be operational by day three of flying operations. Any difficulties encountered in deploying or setting up this complex equipment and making it fully operational effectively delays resupply. Also, if just a single set of testers is deployed to a location, as should happen when only one squadron deploys to an FOL, then the squadron using those testers faces a "single-string" risk, wherein a breakdown of just one tester can halt resupply for an entire group of parts. Resupply shortfalls can result in the decline of aircraft availability below planned levels. "Emergency" setup of an unplanned distribution channel to the FOL

---

[1] See Hosek and Totten (1998).

could mitigate resupply shortfalls resulting from tester-associated problems.

For both the consolidated and the decentralized-no-deployment alternatives that we consider, the need to set up an effective wartime distribution system between repair and operating locations is the major source of risk. Delays in implementation would hinder resupply much as would delays in deploying testers under a decentralized-deployment policy. Similarly, any gap between the order-and-ship time planning assumptions used to plan forward inventory levels and that actually achieved would result in a resupply capability that would be unable to support the planned level of aircraft availability. This risk may increase as customs regulations or the remoteness of operating locations increase.

## SUMMARY OF ALTERNATIVE COMPARISONS

The current decentralized-deployment policy, which calls for slightly higher levels of personnel and testers than those in place today, could provide the same level of support at the same cost as, or at a lower cost than, the alternatives we examine. Disadvantages such as personnel instability, deployment footprint, and equipment setup and "single-string" risks, however, have already led many deploying units to modify their procedures on an ad hoc basis.

Modifying the current system to eliminate AIS deployment, as in the decentralized-no-deployment option, eliminates personnel and equipment deployment requirements but requires a one-time increase in spare parts for the supply pipeline. Furthermore, a moderate level of personnel deployment rather than the elimination thereof may be of most benefit to retention.

The four-FSL/one-CSL option is cost competitive with the current decentralized-deployment option and addresses each of its disadvantages. It offers a moderate level of personnel deployment to nonhostile locations and eliminates equipment deployment and its accompanying risks. These benefits may be somewhat offset by the risk inherent in the need to quickly establish effective wartime intratheater distribution.

## A TEST OF A REGIONAL F-15 REPAIR FSL

During Operation Noble Anvil (ONA), the air war against Serbia, the 48th Component Repair Squadron at RAF Lakenheath implemented the FSL repair concept as part of a system of FSLs set up by United States Air Forces in Europe, thereby formalizing practices they had used on an ad hoc basis for several years.   They were able to successfully support their own aircraft at FOLs as well as concurrent deployments to Southwest Asia using existing assets without any deployment of AIS personnel or equipment.   In fact, between October 1998 and March 1999, as tensions rose or eased, the wing supported by this squadron made seven different partial-unit deployments back and forth from Lakenheath to Southwest Asia and Italy without moving the AIS (Figure S.2).   Normally, Air Force policy would require that these deployments include the AIS, but since all of the units were supported from the Lakenheath FSL, no support equipment had to move.   As a result, airlift requirements for these seven deployments were reduced by 35 C-141 sorties.   More than any theoretical description of the flexibility that FSLs can provide in today's dynamically shifting environment, these operations demonstrated the advantage nondeploying maintenance structures confer in facilitating the repositioning of forces as quickly as political situations change.

The squadron also implemented plans for the Lakenheath avionics maintenance FSL to support an augmentation of F-15s from CONUS for ONA with just half the deployment footprint and personnel that would have been required had the deploying wing's AIS moved to the new FOL.   In a permanent consolidated structure, even this limited deployment of equipment would not have been required because the equipment would already have been in place; thus, only personnel would have had to deploy.   In exchange for the reduction in deployment airlift, the FSL had to rely on a steady flow of transportation to provide resupply to the operating locations.

Lakenheath logisticians used their prior experience, including that gained by the October 1998 deployment to Cervia (shown in Figure S.2), to conduct transportation planning for providing support from an FSL.   This enabled it to provide rapid and responsive resupply of serviceable parts to FOLs from the start of ONA through the intratheater distribution system and a Lakenheath-managed

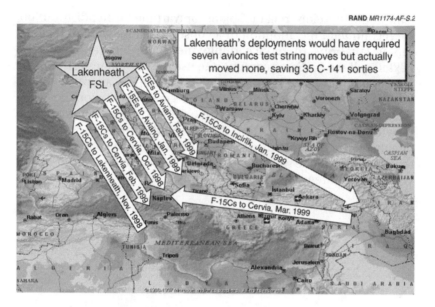

RAND *MR1174-AF-S.2*

**Figure S.2—At Lakenheath FSL Facilitation of Operational Flexibility
During a Time of Heavy Political Turbulence**

"distribution system" that augmented the joint system. The Lakenheath distribution system was critical to the success of the operation.
Other Air Force FSLs established in support of ONA relied solely on
the joint intratheater distribution system but did not find that system
sufficiently responsive.

## CONCLUSION

The key issue in determining whether to adopt an alternative F-15
avionics support structure seems to be the level of risk posed by the
need to quickly establish a wartime theater distribution system. We
recommend that the Air Force review current plans for wartime
theater distribution and then work as part of the joint community to
modify them as necessary to address potential performance gaps.
Even if the Air Force then elects to continue with the current
structure, improving the wartime theater distribution system would
reduce equipment risk. Assuming that the Air Force and joint

community develop "reliable" plans for wartime theater distribution, we recommend the adoption of a consolidated network of regional repair locations to reduce deployment burdens and enhance flexibility if the Air Force continues to use the current testers. Such a network would provide more benefits than ESTS adoption at less cost. If the Air Force proceeds with ESTS implementation, however, the alternative systems would cost more than the current policy and provide fewer benefits. In this case, the reduced personnel deployment requirements and flexibility provided by the alternative structures should be weighed against their associated spare-parts investment requirements.

# ACKNOWLEDGMENTS

Numerous persons inside and outside the Air Force assisted our work. We thank Lieutenant General William Hallin (then AF/IL), for initiating this work and Lieutenant General John Handy (AF/IL), his successor, for continuing the sponsorship and support of this effort.

We have enjoyed support for our research from the Air Force's Major Commands responsible for implementing the Expeditionary Aerospace Force (EAF). Major General Dennis Haines (ACC/LG), Brigadier General Terry Gabreski (USAFE/LG), Colonel William Beechel (ACC/LG), and Brigadier General Donald Wetekam (PACAF/LG) provided access to personnel and data at their operating locations. Colonel Lowman (USAFE/LGX) and Master Sergeant Scott Carter (USAFE/LGX) helped arrange visits to United States Air Forces in Europe bases, facilitated data collection on logistics operations, and provided keen insights into their area of responsibility. Colonel James Daup (ACC/LGS) and Colonel Ralph Austin (USAFE/LGS) sponsored briefings to the Air Combat Command (ACC) and USAFE staff to vet the ideas in this study. At ACC, Lieutenant Colonels Brett Haswell and Alex Cruz-Martinez and Master Sergeant Jose Longoria raised several implementation issues that helped us refine our thinking. Colonel Michael Weitzel (CENTAF/LG) and his staff provided outstanding support for our visits to locations in Southwest Asia. Lieutenant General Stewart Cranston (AFMC/CV) and Colonel Robert Dehnert (AFMC/LG) encouraged us to show how our research could provide insights for the Air Force Materiel Command in implementing the EAF.

At the Air Staff, we thank Major General Michael Zettler and Grover Dunn (AF/ILM), Colonel James Totsch (AF/ILS), Brigadier General Quentin Peterson (AF/ILT), and Susan O'Neal (AF/ILX) and their staffs for their support and critique of this work. In particular, we thank Colonels Jack Leonard, Tom Toole, Gary Sandiford, and Ted Brewer; Lieutenant Colonels Don Virostko, Michael Melendrez, and Joseph Seawell; and Major Craig Romero for providing valuable feedback on earlier versions of this report and for raising several critical issues that led to significant revisions of our analysis. Colonel Sandiford also provided many valuable suggestions on improving the message of our research. We are especially grateful to Colonel John Gunselman for sponsoring our work before senior leaders, even in the knowledge that it would spark a great deal of heated dialogue.

The personnel in the Air Force's F-15 maintenance shops made this analysis possible through their efforts to provide critical data, often from nonstandard sources. We thank Master Sergeant Kenneth Hamlin and Technical Master Sergeant Keith Weisser of the 52nd Component Repair Squadron (CRS) at Spangdahlem; Chief Master Sergeant Frank Levand, Senior Master Sergeant Alan Taylor, Senior Master Sergeant John Powers, and Master Sergeant Robert Shelton of the 48th CRS at Lakenheath; Senior Master Sergeant Eric Johnson, Technical Master Sergeant Todd Anderson, Master Sergeant Michael Brogan, and Master Sergeant Stephen Perkins of the 4th CRS at Seymour Johnson Air Force Base (AFB); and Chief Master Sergeant Gary Dykas and Senior Master Sergeant Mike Shih of the 33rd CRS at Eglin AFB. Lieutenant Colonel Brad Silver, 52nd Chief of Supply, provided many comments on supply-related issues that helped refine our analysis. We thank Lieutenant Colonel Anne Smith, Deputy Commander of the 366th Logistics Group, and Chief Master Sergeant Griffin of the 366th CRS at Mountain Home AFB for their detailed, well-researched comments and corrections. Master Sergeant Tim Flohrschutz from the ACC Systems Office, F-15 Branch, also provided valuable data and key insights that strengthened the work. Finally, we thank Major James Young, Electronic System Test Set (ESTS) Program Manager, for information on ESTS and for the opportunity to gain feedback on our research from the ESTS team.

Our research has been a team effort with the Logistics Management Institute (LMI) and the Air Force Logistics Management Agency (AFLMA). T. J. O'Malley supported LMI participation in this study,

Randy King modeled spare parts requirements for readiness spares packages, and Linda MacArthur extracted Air Force supply system data for our modeling efforts.

AFLMA is a partner in our overall research exploring EAF support alternatives. The continuing support of its staff has been critical to this research. We are especially grateful for the support of Colonel Richard Bereit (AFLMA/CC) and Lieutenant Colonel Mark McConnell (AFLMA/LGM).

Finally, we thank Colonel Rodney Boatright (AF/ILXX) for his encouragement and support. At RAND, Ken Girardini, Don Stevens, Bob Roll, Don Palmer, Craig Moore, Lou Miller, Lionel Galway, and Amatzia Feinberg have helped with reviews and critiques of our work. Gina Sandberg was patient and prompt in preparing the document through numerous versions.

| | |
|---|---|
| AB | Air Base |
| ACC | Air Combat Command |
| ACS | Agile Combat Support |
| AEF | Aerospace Expeditionary Force |
| AFB | Air Force Base |
| AFLMA | Air Force Logistics Management Agency |
| AFMC | Air Force Materiel Command |
| AFWCF | Air Force Working Capital Fund |
| AIS | Avionics intermediate-maintenance shop |
| ANT A | Antenna A |
| ANT B | Antenna B |
| AOR | Area of responsibility |
| ASM | Aircraft Sustainability Model |
| AWP | Awaiting parts |
| BCS | Bench-check serviceable |
| C | CONUS |
| C2 | Command and control |

| | |
|---|---|
| CIRF | Centralized Intermediate Repair Facility |
| CNI | Communications, navigation, and instrumentation |
| CONUS | Continental United States |
| COTS | Commercial off the shelf |
| CRS | Component repair squadron |
| CSIS | Centralized secondary item specification |
| CSL | CONUS support location |
| CSP | Consolidated support package |
| D041 | Secondary item inventory computation system |
| D087 | Weapon system management information system |
| Decen. | Decentralized |
| Decen.(D) | Decentralized-deployment option |
| Decen.(NoD) | Decentralized-no-deployment option |
| DoD | Department of Defense |
| DPG | Defense Planning Guidance |
| DSO | Design support objective |
| EAF | Expeditionary Aerospace Force |
| EARTS | Enhanced Aircraft Radar Test Station |
| EAU | Engine Analyzer Unit |
| EBTS | Electronic Branch Test Set |
| ESTS | Electronic System Test Set |
| FMC | Fully mission-capable |
| FMSE | Fuels mobility support equipment |
| FOL | Forward operating location |

| | |
|---|---|
| FOR | Follow-on operating requirement |
| FSL | Forward support location |
| I&C | Indicators and controls |
| ILM | Intermediate-level maintenance |
| IOR | Initial operating requirement |
| LANTIRN | Low Altitude Navigation and Targeting Infrared Night |
| LMI | Logistics Management Institute |
| LRU | Line-replaceable unit |
| METS | Mobile Electronic Test Set |
| MTW | Major theater war |
| NEA | Northeast Asia |
| NMC | Not mission capable |
| NRTS | Not repairable this station |
| NSN | National stock number |
| OCONUS | Other than CONUS |
| ONA | Operation Noble Anvil |
| OST | Order-and-ship time |
| PAA | Primary authorized aircraft; primary aircraft authorization |
| PACAF | Pacific Air Forces |
| PMC | Partially mission capable |
| POS | Peacetime operating stock |
| PPBS | Planning, programming, and budgeting system |
| PSAB | Prince Sultan Air Base (Saudi Arabia) |

| QOT&E | Qualification operational test and evaluation |
| R | Regional FSL |
| RAF | Royal Air Force |
| RO | Requirements Objective |
| RR | Remove and replace |
| RRR | Remove, repair, and replace |
| RSP | Readiness spares package |
| SBSS | Standard base supply system |
| SPO | System Program Office |
| SRU | Shop-replaceable unit |
| SWA | Southwest Asia |
| TISS | Tactical Electronic Warfare Intermediate Support System |
| USAFE | United States Air Forces in Europe |
| WMP | War and mobilization plan |
| WRM | War reserve materiel |
| WWX | World Wide Express |

# Chapter One

# INTRODUCTION

## EVOLVING OPERATIONAL AND LOGISTICS REQUIREMENTS

The recent increase in frequent, small-scale, short-notice deployments to unpredictable locations has placed a severe strain on Air Force personnel and sustainment capabilities.[1] Political expectations of a continuing need for this high but uncertain operating tempo and rapid response capability have forced the Air Force to develop new concepts of operations. Ad hoc Aerospace Expeditionary Forces (AEFs) have thus evolved into the creation of the Expeditionary Aerospace Force (EAF), which is designed to transform the Air Force into a sustainable fast-strike tool capable of responding anywhere in the world.[2] Consequently, Air Force logisticians are now racing to determine how logistics structures and practices should change to best meet EAF operating demands.

---

[1]Ryan (1998).

[2]The EAF is the "Air Force Vision to organize, train, equip, and sustain itself to provide rapidly responsive, tailored aerospace force for 21st century military operations." Its purpose is to improve response speed and flexibility while decreasing the deployment strain for a CONUS-based Air Force. The EAF will organize the Air Force into ten virtual AEFs comprising combat, mobility, and support resources that Joint Force commanders can tailor to specific missions. Five mobility wings will be paired with two AEFs each and will be on call at the same time as their companion AEFs. The AEFs will operate on a 90-day "on call" window once every 15 months. This will give personnel a more predictable deployment schedule and increase the stability of their personal lives. See Schnaible (1999).

## The Cold War Origins of Today's System

The current Agile Combat Support (ACS) system reflects Cold War demands, and consists of the remnants of a system designed to support one very large conflict in Central Europe and a potentially smaller one in Northeast Asia. During the Cold War, this system had an extensive overseas infrastructure that placed large forces and supplies where they were expected to be employed. For the most part, planning assumed that forces would have to fight with what was already in place, and logistics support policies were designed around this assumption. Since the end of the Cold War, however, political pressure to reduce both defense outlays and overseas forces has led to a smaller force structure based to a greater extent in the Continental United States (CONUS). While the basing structure has evolved, however, the support system has not. Logistics policies have not changed to reflect the fact that most forces today are not where they are likely to fight.

In the past decade, the need for a continuous presence in Southwest Asia (SWA) has highlighted the mismatch between basing structure and support system. Without forces permanently stationed forward in this area, presence has been achieved through a continuous series of deployments. Such deployments have, however, consumed large amounts of resources, interrupted training, and pulled people away from home both frequently and unpredictably, thereby making it difficult to sustain readiness and retain trained personnel.[3] A never-ending string of other contingencies has convinced the Air Force that this has become the rule. Growing evidence of the strain created by these deployments, with their Cold War design-driven practices, led Air Force leaders to develop the EAF and set out to develop new support concepts.

## New Logistics Challenges

The reliance on primarily CONUS-based forces, coupled with new force employment concepts to meet unpredictable political de-

---

[3]See, e.g., Richter (1998b). Note, however, that other research has shown that some deployment may improve retention. For additional reports on the effects of increased operating tempo and personnel tempo on military readiness, see Richter (1998a), Williams (1998), and Hosek and Totten (1998).

mands, presents the Air Force with significant support challenges. The EAF offers a vision in which the Air Force is able to deploy both large and small forces quickly, employ them immediately, and sustain them indefinitely. To meet the demanding EAF timelines for immediate employment and sustainment, units must be able to deploy rapidly to forward operating locations (FOLs) and quickly establish logistics processes.

The need to deploy large forces quickly is driving the Air Force to find ways to reduce the support resources that must be shipped during initial combat operations so that more combat forces can deploy during this period. Cutting deployment requirements for support resources can be accomplished in two ways. The first is to pre-position resources where they are most likely to be needed. The second is to eliminate the need for certain resources at forward locations either by eliminating the need for the resource altogether or to change practices to make "reachback" effective. Limited access to foreign bases, uncertainty about contingency locations, resource constraints, and difficulty in protecting forward locations favor the latter option. In the long term, new aircraft designs promise to reduce maintenance requirements, precision-guided munitions promise to decrease the weight and cube of munitions, and more effective weapon systems promise to decrease needed force sizes. However, dramatic design-driven changes will occur gradually over an extensive period. In the meantime, it is therefore imperative that the Air Force change structures and processes in ways that reduce the deployment burden. Such innovations will also enhance the impact of technological developments.

Reducing the "footprint" (i.e., materials needed) for support by supplying operations through responsive resupply has thus become an attractive method of operations. With a responsive resupply pipeline, for example, substantial early-airlift capacity devoted to the deployment of bulky maintenance equipment for component repair at FOLs could be traded for constant and much smaller capacity dedicated to spare-parts airlift through the duration of a conflict. And this airlift, depending on the resources available at the FOL, could be even less than that normally needed to sustain forward-deployed maintenance personnel. This line of reasoning can be found in a growing body of Department of Defense (DoD)-wide work examining distribution-based support structures that aim to keep as much materiel and equipment as possible away from the "front lines" in efforts to increase response speed.

Maintaining readiness to meet potential major theater war (MTW) requirements while temporarily deploying a significant portion of the force to meet "boiling peacetime commitments" presents additional support challenges.[4] The support system must be able to accommodate EAF operations in a variety of locations with different infrastructures in any area of responsibility (AOR). At the same time, it must be flexible enough to deal with dynamically changing events and capable of shifting rapidly from one kind of operation to another.

## THE AGILE COMBAT SUPPORT SYSTEM OF THE FUTURE

Recognizing that future operational capabilities depend to a large degree on ACS infrastructure and process decisions made now, Air Force leaders have begun to examine how the combat support system can best meet the EAF's logistics challenges. Efforts are thus being made to develop alternatives and then evaluate how well these alternatives support EAF operational objectives in comparison to the current system. At the same time, the Air Force recognizes that today's budget environment will require that future support structure and policy alternatives cost about the same as today's, if not less.

### A New ACS Concept That Mixes CONUS and Forward-Based Assets

Since the initial creation of ad hoc AEFs, RAND researchers have been examining current and alternative support concepts in efforts to understand how well they support expeditionary operating concepts.[5] The starting point for this research was the initial AEF "line in the sand," a declaration stating that the Air Force would develop a force capable of dropping bombs on targets anywhere in the world within 48 hours. Using simple spreadsheet models to evaluate sup-

---

[4]"Boiling peacetime commitments" is a term coined by General John Jumper to describe the requirements to deploy a significant amount of aerospace forces during peacetime to ensure global stability.

[5]Tripp et al. (1999).

ply system processes to support a variety of commodities, researchers found that for some commodities the current system could not achieve this target, as it just took too long to move and set up equipment from home station bases. The use of forward prepositioning, however, was found to make 48-hour employment achievable. At the same time, prepositioning increases resource requirements and thus boosts costs, so it is not feasible everywhere.

Research was then conducted to explore the actual capabilities of the current process and the extent to which these capabilities could be improved through policy and structure modification. For which commodities would a distribution-based system be appropriate today? Which would need prepositioning to meet EAF goals? And are there options that combine elements of the two to provide a better overall solution?

We briefly review one such commodity analysis—for heavy bombs—that helped illuminate the research issues and from which a new vision of a possible future Air Force ACS began to emerge. Comparing spin-up time, or the time needed to launch sustained combat operations, to costs incurred by prepositioning materiel is one of the tradeoffs posed by alternative ACS structures. Analysis revealed that heavy bombs must be positioned at FOLs in order to achieve a 48-hour employment goal (see Figure 1.1). Even with FOL prepositioning of these munitions, current processes can barely achieve that aggressive timeline. However, the right bar and line in Figure 1.1 show that while stockpiling munitions centrally in CONUS is much cheaper, spin-up time is twice as long and exhibits more variability. But what if these factors were balanced? The middle portion of Figure 1.1 displays the spin-up time and costs of a system in which prepositioned munitions are consolidated at a few strategic locations to accommodate the possibility of AEF operations in multiple AORs. This range of alternatives illustrates the tradeoffs between munitions costs and spin-up time generated by varying the ACS infrastructure. Deciding which ACS structure to use depends on the relative values the Air Forces places on operating costs and spin-up time as well as other logistics support metrics.[6]

---

[6]For more information on the nature of these tradeoffs, see Galway et al. (2000).

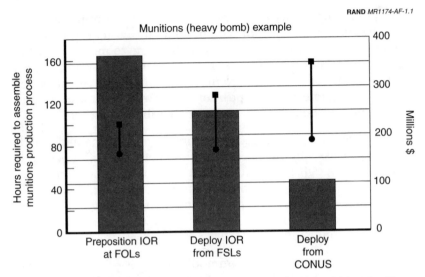

RAND *MR1174-AF-1.1*

NOTE: IOR (initial operating requirement) is the amount of materiel that should be in place to allow for sustained combat operations from the commencement of combat until resupply.

**Figure 1.1—Tradeoff Between Investment/Recurring Cost and Spin-Up Time (bars represent cost and lines represent time)**

This munitions example and similar analyses of other commodities have indicated that the peacetime location of ACS assets and processes—whether at FOLs, forward support locations (FSLs) in or out of theater, or CONUS support locations (CSLs)—can be a critical element in the tradeoff between cost and spin-up time as well as in tradeoffs between many other performance measures important to EAF operational concepts. These metrics include spin-up time, cost, deployment airlift footprint, operational risk, operational flexibility and personnel turbulence.

A potential model for future alternative ACS systems has begun to emerge from individual commodity analyses evaluated using these metrics.[7] In this model, each ACS option is based on the peacetime positioning of assets and processes and on where those assets will

_____
[7]Galway et al. (2000).

move in response to aircraft deployments. For each commodity and process, positioning should begin with an evaluation of options using the six metrics, and a selection should then be based on the relative values the Air Force places on each measure of merit. An ACS system capable of handling a wide spectrum of operations is likely to be built from a combination of five components:

1. **FOLs.** Some FOLs in critical areas would require that substantial resources be prepositioned to allow for rapid deployments of heavy combat aerospace packages, perhaps augmented with more austere FOLs that would take longer to spin up. In parts of the world where quick response is not necessary or missions are not demanding, FOLs might all be of the second form.

2. **FSLs.** The configuration and specific functions of FSLs would depend on their geographic location, the threat, and the costs and benefits of using current facilities. Western and Central Europe, for example, are currently stable and secure, and an FSL established there could support operations in adjacent AORs such as SWA.

3. **CSLs (either decentralized or consolidated).** Analogous to the FSLs, a similar support structure is needed to support CONUS forces, since some repair capability and other activities may be removed from units. As with FSLs, a variety of different activities may be set up at major Air Force bases, convenient civilian transportation hubs, or Air Force or other defense repair depots.

4. **A transportation network connecting the FOLs and FSLs with each other and with CONUS, including en route tanker support.** This is essential, as FSLs and FOLs must be connected via transportation links. FSLs themselves might even be transportation hubs.

5. **A logistics command-and-control (C2) system.** This would coordinate the system, organize transport and support activities, and allow the system to react swiftly to rapidly changing circumstances.

The appropriate balance between these building blocks depends on numerous factors. The primary focus of the system should be on areas of vital U.S. interests, e.g., on clusters of FSLs and FOLs in Korea,

SWA, and Europe. In some areas, the threat may not be time-critical (e.g., it may not include armored attack over an unobstructed corridor), in which case it may be acceptable for FOLs to be relatively undeveloped, perhaps with potential locations merely identified, rather than fully developed with substantial prepositioning. Actual AEF operating and support locations will depend on variables such as existing infrastructure and force protection, political factors (especially those affecting access to bases and resources), and the potential that locations offer for building relationships with allies and host nations.

Near-term ACS structure design decisions should consider current forces and support processes. As new policies and practices are discovered and implemented, as the Air Force gains experience with expeditionary operations, and as new technologies for ground support, munitions, shelter, and other resources become available, decisions will need to be revisited and the support system configuration adjusted to reflect new capabilities. Transportation time improvements, for example, could render support operations from CONUS more effective and allow for reductions in prepositioned resources.

## Implications of This New ACS Concept for F-15 Avionics Maintenance

F-15 operations are currently supported by a three-echelon maintenance system. At the flight line, mechanics remove and replace major components using stock kept in base supply. The second echelon, intermediate-level maintenance (ILM), consists of a component repair activity in the logistics group of each F-15 wing that is collocated with the aircraft squadrons. ILM serves to replenish base supply for a designated set of components by using "piece parts" stored in base supply to repair these components. Current policy calls for F-15 intermediate avionics maintenance activities to deploy with aircraft to FOLs. The third echelon is depot repair, which handles components that intermediate maintenance is unable to repair as well as those designated for depot repair only. Rather than rely on distribution for support, this system depends on moving mass to the FOL, maintenance equipment, personnel, and spare parts.

This report examines how alternative F-15 avionics intermediate-maintenance structures based on distribution, derived from the new ACS concepts described in the previous section, would impact the cost-effectiveness of F-15 avionics support. The analysis focuses on understanding the implications, benefits, and affordability of conducting intermediate-level component maintenance at FSLs and CSLs instead of at home bases and FOLs, as is the policy today. The analysis considers the support requirements necessary to meet peacetime and wartime demands for all avionics components for all F-15Cs and F-15Es.[8] In addition to examining the implications of maintenance locations, this research presents an example of how technology investment compares with policy options such as consolidating maintenance to achieve EAF objectives such as reducing the deployment footprint. Specifically, this report compares how investments in a new, downsized avionics tester that would deploy with F-15 squadrons would compare with the continued application of existing testers used instead at consolidated repair locations.

## ANALYSIS APPROACH

### ACS General Analytic Approach

We use an employment-driven modeling approach developed in earlier RAND research for planning and evaluating ACS structures.[9] The first step in this approach is to determine the force packages (i.e., aircraft numbers and types) and operating profile necessary to accomplish anticipated missions (see Figure 1.2). The second step consists of two elements, the first of which involves translating this information to the operating demand on the logistics system—in this case the expected number of avionics components that must be replaced on each F-15 aircraft each day. Using as their basis the support structure (the desired mix of FOLs with prepositioned materiel,

---

[8]We apply specific definitions to the EAF ACS elements with regard to F-15 avionics intermediate maintenance. CSLs represent consolidated repair activities in CONUS, and FSLs serve as forward consolidated repair activities. In some cases FOLs will simply be forward operating bases and in others they will have wartime repair capability (but not prepositioned).

[9]Tripp et al. (1999).

FSLs, and CSLs) and policies, logistics models then determine the allocation of resources needed to meet these demands (Figure 1.2).

The third step is to evaluate how each structure and its resource requirements affect ACS metrics. The ACS metrics in this general framework consist of the operationally derived metrics of spin-up time, deployment airlift footprint, costs, risk, and flexibility. If the metric evaluation reveals that the alternatives do not meet EAF needs, possible revisions of operational objectives can be evaluated or alternative support practices or technologies considered for development and evaluation.

Added to the ACS metrics for this analysis is the EAF goal of reducing personnel turbulence. The continuous cycle of deployments in recent years to meet boiling peacetime commitments—including repeated conflicts with Iraq—has increased strain on personnel, which is believed to have led in turn to problems in retaining skilled aircraft technicians. The Air Force EAF implementation plan attempts to reduce personnel turbulence by placing deployments on a predictable schedule. In the plan, however, people will still be exposed to potential deployments, sometimes to hostile locations, every 15

RAND *MR1174-AF-1.2*

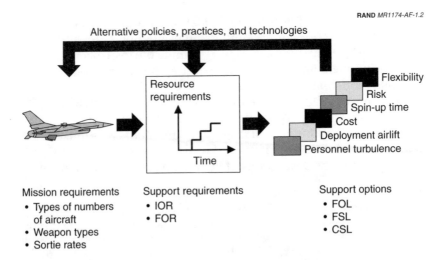

Figure 1.2—A Standard Approach for Evaluating ACS Systems

months—and while this schedule should reduce deployment "surprises" and better balance deployment demands among units, individual personnel might still be exposed to turbulence. The effect that support structures might have on personnel strain by reducing deployment demands is thus another benefit considered.

The purpose of this analysis is not to combine these metrics into one "score"; rather, it is to show how alternatives are likely to affect the metrics. The Air Force must then determine how much value it places on each of the goals and which alternative presents the best set of tradeoffs. In the report, we quantify the cost and deployment footprint metrics while conducting more qualitative reviews of the other metrics.

Where feasible and appropriate, we relied on existing models, with modification as required, to determine mission and support requirements, creating entirely new models only where necessary.

## Evaluations Across a Spectrum of Operational Requirements

The peacetime and wartime locations of intermediate-maintenance assets comprise the basis of the alternative F-15 support structure designs evaluated. For each structure, the set of locations, in conjunction with support policies, creates a unique requirement for the five resources that complete the definition of the structure. These resources are:

- intermediate test stands and fixtures
- personnel
- spare parts
- transportation
- infrastructure (a combination of facility requirements and the permanent home stations of personnel in this analysis).

The objective of this analysis is to determine the costs and benefits of alternative F-15 avionics maintenance structures that can satisfy the entire spectrum of operational requirements. These include the Defense Planning Guidance's (DPG's) two-MTW scenario, small-scale AEFs, and boiling peacetime operations. Also considered are support system goals for rapidly deploying, immediately employing, and indefinitely sustaining forces to meet two MTWs.

The total demand placed on a support location in terms of avionics spare parts needed per day, whether that location is collocated with flying units or not, is a product of the number of aircraft being supported and their operating profile (sortie rates and flying hours). In the structures analyzed, the maximum daily surge and sustained demands for avionics spare parts at both the individual-location and total worldwide system levels are produced by the two-MTW scenario. Thus, the total system and each location must have a set of support resources that will meet the two-MTW spare-parts demand, and this set of resources should meet the demands of the other missions as well.

The mix of maintenance, supply, and transportation resources that each structure requires affects the desirability of the structure for supporting other types of missions. Even though two structures can support the same demand rate at an FOL, for example, one structure might do so with a substantially lower deployment airlift requirement and might thus be favored for rapid-response missions.

## OBJECTIVE AND ORGANIZATION OF REPORT

Many today are quick to call for distribution-based systems that support operations from existing infrastructure locations, primarily within CONUS. The key argument in favor of this change pivots on a reduction in deployment airlift requirements. Those who oppose such a change typically do so because they do not believe the existing Air Force distribution and DoD processes and infrastructure could properly support mission requirements. The objective of this report is to bring analysis to this debate. What benefits would result from moving to a distribution-based system, be it one that relies on existing infrastructure or an alternative based on consolidation and some forward positioning? Would these systems provide the proper level of support? Or, perhaps more important, would unaffordable increases in resources be necessary to enable these systems to provide the required level of support?

The remainder of the report is organized as follows:  Chapter Two details the alternative structures, logistics policies, and technology options that will be evaluated.  Chapter Three describes the models developed to determine the resource requirements of each alternative structure, and Chapter Four presents the results of the models. Together, these chapters define the policies and resource requirements of the alternative structures in order to evaluate how they affect the EAF goals discussed in Chapter Five.  Chapter Six considers improvements that might be possible with the consolidated structures, and Chapter Seven concludes with a summary of the advantages and disadvantages of the support structures and recommendations for the Air Force.

# Chapter Two

# SUPPORT STRUCTURE OPTIONS AND THE
# DECISION SPACE

F-15 avionics component demands can be met through several support structures, all of which consist of combinations of maintenance and supply policies. Combining a new avionics tester development program with ongoing Air Force efforts to improve the distribution system creates a large, complex decision space.

Coupling concepts for maintenance and spare-parts supply creates the underlying structures of alternative support options. Maintenance concepts are in turn derived from two design factors: first, whether ILM is collocated with supported aircraft; and second, the level of maintenance consolidation. In the Air Force, spare-parts planning is done separately for wartime readiness spares packages (RSPs) and peacetime operating stock (POS).[1] Resupply start time,

---

[1]The Air Force sets local inventory levels separately for peacetime and wartime and segregates the two sets of stock. The inventory to support peacetime, home-base operations, called peacetime operating stock (POS), supports all of the squadrons at a base through consolidated inventory; inventory levels are based on the demands of peacetime flying profiles. The inventory to support wartime operations (used for all deployments) is set aside in readiness spares packages (RSPs) that are packed and ready for immediate deployment. Each squadron of aircraft has a dedicated RSP based on its projected wartime flying profile. The on-hand level for RSPs is required to be at the same level as the inventory requirements objective (RO) upon deployment. This follows from an assumption that resupply demands placed on the system prior to deployment will not be rerouted to the deploying location, meaning that when deployment occurs, the resupply pipeline is effectively empty. Since it is possible for all of the POS's inventory position to be due in rather than on hand at the time of deployment, it is necessary to assume no use of POS inventory in the RSP. Also, it would be extremely difficult to divide a consolidated inventory package that supports multiple squadrons at home into separate sets of stocks upon deployment. Further, the RSP

which is determined by how long it is expected for the wartime sustainment distribution system to begin operating, and the design support objective (DSO), which is defined as the percentage of aircraft that must be operational on a specified number of days after the start of combat, are key input parameters in the RSP planning process. Maintenance location influences both RSP and POS planning by affecting the pipeline length between the repair activity and flightline supply points, which in turn affects both inventory operating and safety stock levels. The length of the spare-parts resupply pipeline is also affected by the capabilities of the distribution system—which, in conjunction with repair time, controls the operating level of parts in the pipeline. Together, the distribution system's capabilities, measured by order-and-ship time (OST), combine with maintenance locations to determine resupply pipelines lengths for each structure.[2] In light of continuing DoD and Air Force efforts to improve distribution system performance, OST is examined as a variable across a wide spectrum of performance levels to better understand the sensitivity of spare-parts requirements to OST across a range of support structure options. Understanding the possible implications of better OSTs would be useful in making balanced cost-benefit decisions about possible OST improvement efforts.

Currently, the Air Force is developing a new F-15 avionics tester to replace several testers currently in use. Because some uncertainty exists about the development of the new tester—the Electronic System Test Set (ESTS)—we also examine how tester choice affects the costs of different support structures. This may provide some general insight into how policy decisions can compare with technology options in the pursuit of goals.

As previously outlined, five major factors affect the resource requirements of the support structures modeled in this analysis: repair location, level of repair consolidation, test equipment, spare-parts

---

quantities are higher because they are designed to support higher wartime demand rates, and separating inventory by squadron increases the overall requirement (i.e., POS takes advantage of the economies of scale offered by combining inventory for all squadrons at a home base).

[2]Order-and-ship time is the duration from the creation of a requisition by base supply to the completion of receipt processing by base supply, assuming that the item is ready for shipment at the source of supply at the time the requisition is received (i.e., no back-ordered requisitions).

planning parameters, and distribution system capability or OST (Figure 2.1).  Of the 200 possible combinations of the values we considered for these factors, we concentrate in this report on 36 that provide insights into the effects of varying each factor through the range of possibilities.  In the remainder of this chapter, we describe these 36 structures by detailing six support structure options (consisting of combinations of wartime repair location, level of repair consolidation, and spares planning parameters), three sets of OST assumptions, and finally two possible sets of avionics testers that together generate the 36 points in the decision space of this analysis.

RAND *MR1174-AF-2.1*

Wartime repair location
- Forward (AIS deploys)
- Rear (AIS does not deploy)

Repair consolidation
- Decentralized
- Consolidated
  - One, three, four, and five locations

Test equipment
- Current testers
- ESTS

Spares planning parameters
- War reserve material (WRM)
  - DSO
  - Resupply start time
- Stock locations
  - Local
  - Consolidated

OSTs
- Current
- Fast
- Faster

Figure 2.1—Five Key Factors Affecting Analysis Results

## SUPPORT STRUCTURES

The basis of the six support structures considered in this analysis are a continuum of consolidation options ranging from decentralized maintenance, in which each unit has its own ILM capability, to consolidated maintenance, where maintenance is performed at only one site in CONUS.  Pairing consolidation options with one of the two levels in the deployment factor—i.e., deploying or not deploying ILM capability—completes the maintenance concept of each support structure.  Completing each structure is an associated set of supply policies.

We assume that all of the structures are supported by the existing Air Force depot system, which repairs, procures, and stores spare parts.

The Air Force's current avionics support structure for F-15 avionics, which completely decentralizes maintenance and deploys avionics intermediate-maintenance shops (AISs) to FOLs, serves as the baseline for our comparison. Under current policy, ILM equipment is located at each home operating base and deploys with aircraft to FOLs. Table 2.1 displays these six structures, showing the maintenance concepts on the left and their corresponding spare-parts concepts on the right. The remainder of this section describes these structures in detail.

## The Current Decentralized System

In the current decentralized system, termed "decentralized deployment" and summarized in the first row of Table 2.1, each peacetime F-15 operating base has ILM personnel and avionics test equipment that deploy in test string sets (one full set of each tester type is called a test string) to FOLs with the F-15s.[3] RSP planning in this system assumes that the AIS will be operational within three days of deployment and that resupply will begin on day 30 of combat operations. Therefore, RSPs must have enough spare line-replaceable units (LRUs) to satisfy demands for the following requirements:

- remove-repair-and-replace (RRR) LRUs to cover the demand over the length of time it takes to repair LRUs at the AIS

- additional RRR LRUs to cover the initial three-day period during which the AIS will not be operational

- RRR LRUs to accommodate the expected number of LRUs found not to be repairable during the first 30 days

- remove-and-replace (RR) LRUs to cover demand over the 30 days it takes to establish resupply

---

[3]The number of test strings per base is based on the number of combat-coded (those squadrons designated for operational combat missions) and training primary authorized aircraft (PAA) at the base. In most but not all cases, this results in one test string per squadron.

**Table 2.1**

**Summary of Maintenance and Spare-Parts Structures**

| Structure | F-15 Avionics Maintenance Option | Associated Spares Concept |
|---|---|---|
| Decentralized deployment | Deploy repair capability with flying units | Current RSPs (remove and replace [RR], remove, repair, and replace [RRR]), resupply starts day 30, AIS operational day 3) Two echelons of supply (base/depot) |
| Decentralized no deployment | Provide repair from home bases | Retain current RSP DSOs; convert AIS items to RR; provide time-definite resupply beginning day one; and adjust for bench-check serviceable (BCS) items (option one) |
| Consolidated maintenance — four FSLs + CSL — three FSLs + CSL — two FSLs + CSL — one CSL | Provide repair from consolidated locations — Pacific Air Forces (PACAF), United States Air Forces in Europe (USAFE), SWA, CONUS — PACAF, USAFE, SWA, CONUS — PACAF, USAFE, CONUS — one CSL only | Deploy new RSPs for MTW and AEF deployments — Raise DSO and provide spares for seven days with FSLs and ten days with CSL only; provide consolidated support package (CSP); provide time-definite resupply beginning on day one and adjust for BCS (option two) — "2.5" echelons of supply — Could also use RSP option one |

- sufficient shop-replaceable units (SRUs) to repair RRR LRUs for 30 days of operations.[4]

The demand for RSP calculations is based on the expected number of removals confirmed as unserviceable that should result from the ex-

---

[4]The Air Force designates F-15 LRUs as either two-level or three-level items. By policy, two-level items should be repaired only at depots, bypassing the AIS, while three-level items are tested and repaired at the AIS and sent back to the depot only when found not repairable at the AIS. Two-level items are designated RR items, and three-level items RRR.

pected wartime flying profile of aircraft, which is expressed in terms of flying hours and sortie rates.[5]  War-planning factors specify two rates: a surge rate for the first seven days of operations and an operational sustainment rate.  Each squadron has a dedicated RSP.

POS is calculated for each base, pooling the demands of all squadrons on the base.  As with RSPs, POS calculations are based on local repair of RRR LRUs, which requires it to have:

- RRR LRUs to cover the demand over the length of time it takes to repair LRUs at the AIS

- SRUs to repair these RRR LRUs

- RRR LRUs to accommodate the demand resupply pipeline length from the depot for LRUs found to be not repairable on base

- RR LRUs to cover the pipeline between depots and bases.

The demand for POS is based on the actual demand experienced at the home bases during peacetime training operations.

### Eliminating Maintenance Deployment from the Decentralized Structure

The second ACS alternative (decentralized, no deployment), summarized in the second row of Table 2.1, is generated by eliminating AIS deployment from the current structure.  This removes SRUs from deploying RSPs and adds the transit time between home bases and FOLs for both carcasses and serviceable LRUs to the pipeline time for RRR LRUs.  The SRUs currently in the RSPs would constitute a non-deploying RSP kept at home bases.  We assume that under this structure resupply starts immediately (the resupply start time is the day on which order placement and processing begins), because covering 30 days of demand for RRR LRUs with stock only would be prohibitively expensive.  It is also necessary to adjust the stockage level for so-called bench-check-serviceable (BCS) items, because these items will not be identified as such until they are sent back to

---

[5]Current RSP inventory requirements do not include levels to accommodate removals that result in no deficiency found.  It is assumed that screening will quickly identify such LRUs and that they will then be returned to stock for reissue.

home operating bases for testing.[6]   The amount of POS would be the
same as for today's decentralized-deployment structure, since there
would be no changes in peacetime structure and practices.

## Consolidated Maintenance Structures

The opposite of the current decentralized structure would involve
consolidating all intermediate maintenance at one location within
CONUS to support all peacetime and wartime missions.  Employing
varying numbers of FSLs with consolidated regional repair capability
in conjunction with one CSL spans the spectrum of possibilities between full consolidation and decentralization.  For this analysis, we
consider three such structures between the consolidated and decentralized extremes.  The last row of Table 2.1 summarizes the four resulting consolidated structures, which consist of a CSL and 0, 2, 3, or
4 FSLs.

The first of the three intermediate options employs three support locations:

- a CSL to support peacetime and combat operations in the
  Americas

- an FSL in Europe to support SWA, European, and African operations

- an FSL in the Pacific to support operations in East Asia.

Central Asia would have to be supported from either the Pacific or
the European FSLs.

Our analysis collocates the FSLs with the F-15 bases at Royal Air
Force (RAF) Base Lakenheath and Kadena Air Base (AB).  We model
Seymour Johnson Air Force Base (AFB) as the CSL, although a depot

---

[6]LRUs removed from an aircraft and sent to the AIS that show no evidence of a fault
on the test station are designated BCS and are returned to supply for reissue.  They are
not included in the Air Force's demand history data, and thus stock to cover these removals is not included in POS or RSP computations.  Without testers to detect BCS
conditions, all LRUs will have to be shipped to the repair location, triggering full use of
a lengthy pipeline.

or another base could be used.[7]  The two other intermediate options include a three-FSL-plus-CSL configuration, which adds an FSL in SWA, and a four-FSL-plus-CSL configuration, which further adds an FSL at Elmendorf.  Figure 2.2 shows the FSL/CSL locations for each structure.

In each consolidated option, repairs would occur at FSLs and at the CSL.  Each FSL would support operations in its area ranging from peacetime commitments to an MTW.  This means that FSLs would have to be designed to handle MTW demand, which would give them

RAND *MR1174-AF-2.2*

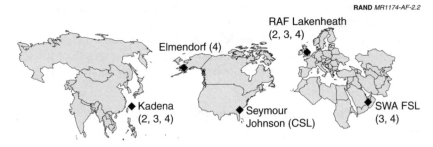

◆ Consolidated support location (numbers indicate the structures in which the location serves as an FSL, e.g., "2" represents two-FSL plus one-CSL case)

**Figure 2.2— FSL and CSL Locations for Each Structure**

---

[7]Seymour Johnson AFB was selected as the CSL in this analysis because it has the most F-15s of any base (two combat-coded squadrons and two training squadrons of F-15Es for a total of 96 PAA).  Each aircraft collocated with the CSL reduces the operating stock component of the worldwide POS, because it can be resupplied via a short, local repair pipeline rather than relying on the combined repair and distribution pipeline between the base and the CSL.  Moving the CSL to a depot would similarly increase the POS requirement at Seymour Johnson owing to the increased pipeline requirement.  However, this increase would be offset to some extent by the reduction in pipeline length for replenishing SRUs at the CSL, since they could be stored at the designated CSL depot.

Another issue would involve management control.  Moving the CSL to a depot would likely mean that the CSL would become part of the Air Force Materiel Command (AFMC), whereas keeping it at a base would allow it to remain under the command of the Air Combat Command (ACC).  The issue is whether the combat forces would feel comfortable giving up control over ILM and whether AFMC would have the same operational and evaluation incentives to ensure a CSL with performance as responsive to ACC's needs as one controlled by ACC.

excess capacity during peacetime. When operating with FSLs, the CSL is designed to meet the peacetime needs of all CONUS-based aircraft and will thus have excess capacity when CONUS units are deployed to other theaters for contingencies and MTWs. When operating without FSLs—the CONUS-only support option—the CSL is designed to meet a two-MTW demand.

To improve the efficiency of maintenance operations, we modify the consolidated concept to require a shift of excess labor capacity from the CSL during wartime to bring FSLs up to full MTW support resource requirements. The Air Force already has enough testers to simultaneously meet the maximum requirements at all proposed FSLs (maximum requirement during an MTW) and at the CSL (maximum requirement when no aircraft are deployed from CONUS); therefore, setting the peacetime tester capacity to the MTW requirement at FSLs would not produce any new acquisition costs.[8] In the event of an MTW, a small number of CSL personnel could augment FSL capabilities, thereby minimizing personnel requirements. Because no equipment needs to move, the deployment footprint becomes minimal. Additionally, the FSLs proposed in this report would have sufficient peacetime resources to handle projected AEF and boiling peacetime demands without requiring personnel augmentation.

As with the decentralized-no-deployment option, flight-line maintenance personnel would conduct RR maintenance and send all LRUs to assigned repair sites. Again, we assume that resupply would start on day one. This would require assured resupply, with resupply requests beginning on day one of the deployed operation.

The consolidated supply policy we assessed exploits consolidation for both supply and repair by employing a consolidated support package (CSP) at the FSLs and CSL. In effect, the CSP is a pooled version of RSPs whose serviceable LRUs can immediately satisfy requisitions from deployed units without having to wait for repair

---

[8]For example, a two-FSL/one-CSL structure might need three of tester "X" at each FSL and eight at the CSL to support peacetime operations along with eight at each FSL and two at the CSL to support two MTWs. As long as the Air Force already has 24 "X" testers, it could permanently keep eight at each of the three locations without incurring additional investment cost.

completion. The CSP would also contain SRUs that the FSLs and CSL would use in LRU repair. The CSP should provide a buffer of serviceable stocks, giving deployed units confidence that stock will be available to meet their needs, including RSP replenishment. Unserviceable LRUs sent back to FSLs or the CSL, once repaired, would refill the CSPs.

## RANGE OF OST ASSUMPTIONS

We compare the six support structures over three different peacetime OST assumptions (Figure 2.3). For our baseline, OSTs within a region of CONUS are set at seven days and OSTs between CONUS and other-than-CONUS (OCONUS) locations at ten days, based roughly on current OST values used to compute F-15 avionics inventory levels.[9] We assume retrograde times for unserviceable reparables to be equal to OSTs over similar routes, so total pipeline time for the retrograde, repair, and return of an LRU becomes the repair time

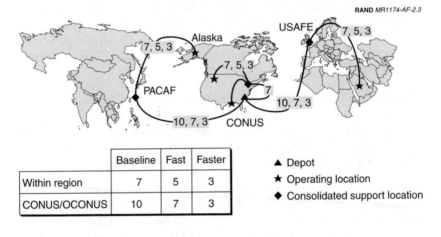

| | Baseline | Fast | Faster |
|---|---|---|---|
| Within region | 7 | 5 | 3 |
| CONUS/OCONUS | 10 | 7 | 3 |

▲ Depot
★ Operating location
◆ Consolidated support location

**Figure 2.3—OST Assumptions**

[9]The average OST for F-15 items in the March 1999 Recoverable Item Computation System (D041) database was nine days.

plus double the OST.[10]   We apply the baseline peacetime OSTs under wartime conditions for all cases.

Analysis of the Air Force World Wide Express (WWX) contract and discussions with Air Force logisticians led us to consider "fast" OSTs for peacetime, whose assumptions are five days within a region or within CONUS and seven days between CONUS and OCONUS locations.[11]   Improvements to the current delivery system seem likely to render fast OSTs feasible in the near to midterm.

Some in the Air Force aim for still faster OSTs in the belief that they should be feasible in the distant future.  Whether feasible or not, consideration of even faster times is of interest in that it lends insight into the worldwide logistics system that may prove of benefit in evaluating the cost-effectiveness of further improvements.  If radically better OSTs are in fact shown to produce dramatic and  quantifiable benefits, those working to improve the system will be given an added incentive to succeed.  The faster OST assumptions are three days for all locations.

## TESTER CONFIGURATION OPTIONS

The current life-cycle status of avionics intermediate testers adds the choice of test string configuration to the trade space (a test string is one tester of each test station type forming a complete set).  Currently there are two different test string configurations, one for F-15Cs and one for F-15Es.  Both configurations are scheduled to change over the next three years as the Air Force fields the ESTS, although significant uncertainty remains about the fielding and ca-

---

[10]Retrograde time is defined as the time it takes to remove an unserviceable item from an aircraft or an LRU and ship it to its repair destination.

[11]See Contract No. F11626-98-D-0031 (DHL Worldwide Express), 1 October 1998 and Contract No. F11626-98-D-0032 (Federal Express Corporation), 1 October 1998. For most OCONUS locations, the contracts guarantee three days, not including the day of pickup, weekends, or holidays. Adding one day for receipt take-up after arrival at base supply results in a range of five to eight days (for a three-day weekend situation), assuming shipments take the full three days, and most should be seven or less.

pabilities of that system.[12] We include both the current tester configurations and the ESTS configuration in our analysis.

The F-15C configuration consists of Antenna A and Antenna B to test the radar antennas, the Tactical Electronic Warfare Intermediate Support System (TISS) to test electronic warfare LRUs, the Engine Analyzer Unit (EAU), and five avionics testers: displays; microwave; computers; communications, navigation, and instrumentation (CNI); and indicators and controls (I&C). The Mobile Electronic Test Set (METS) configuration used to support F-15Es replaced three of the five avionics testers: computers, CNI, and I&C. The METS, however, tests only a subset of the LRUs that were formerly tested on the three testers it replaced. From a deployment standpoint, it also offers the advantage of being substantially smaller than its predecessors. The F-15E shops have also received the Enhanced Aircraft Radar Test Station (EARTS) to replace the antenna test stations. The EARTS is more automated, which is intended to improve troubleshooting capability and consistency.[13] Both the METS and the EARTS are compatible with F-15C LRUs. The ESTS will replace the five original avionics testers or the METS as appropriate and will operate in conjunction with the antenna testers, the TISS, and the EAU. Figure 2.4 compares the three test string configurations; Figure 2.5 illustrates examples of the testers.

There are two reasons to analyze the support structures using different tester configurations. First, by comparing the options under each tester configuration, such an evaluation serves as a form of sensitivity analysis; if the same option looks best regardless of its configuration, confidence in it should increase and the Air Force can decide on the support structure regardless of ESTS implementation. Second, the analysis can inform ESTS program decisions and, more

---

[12]By 2001, the Air Force aims to field to all shops a new, downsized tester, the ESTS, to replace the entire series of microwave, displays, computers, CNI, I&C, and METS test stations. The ESTS recently failed its qualification operational test and evaluation (QOT&E), however, so its delivery and possibly even its fielding are not certain. See "F-15 ESTS Beddown Schedule" (1998).

[13]Automating a test station is akin to error- and mistake-proofing in commercial industry. By automating the test sequence, the Air Force reduces the possibility that the technician will make a mistake in the test procedure. It also enforces a standardized process that uses all of the expert knowledge available.

RAND *MR1174-AF-2.4*

| Current Inventory | | ESTS plan F-15 ESTS |
|---|---|---|
| Original fielding F-15C AIS | METS fielding F-15E AIS | ESTS plan F-15 ESTS |
| ANT A ANT B | EARTS | ANT A and B/EARTS |
| TISS | TISS | TISS |
| EAU | EAU | EAU |
| Displays | Displays | ESTS |
| Microwave | Microwave | |
| Computers I&C CNI | METS (F-15 C-compatible) | |

**Figure 2.4—F-15 Avionics Tester Configurations**

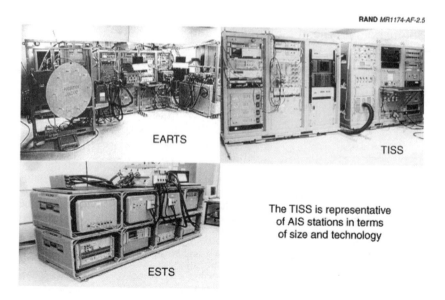

RAND *MR1174-AF-2.5*

EARTS

TISS

ESTS

The TISS is representative of AIS stations in terms of size and technology

**Figure 2.5—F-15 Avionics Testers**

important, can demonstrate a way to expand the trade space for future support decisions regarding other components and aircraft.

Adding the ESTS allows us to expand the trade space not only by comparing different types of technology but also by comparing solutions employing existing policy (i.e., decentralized maintenance) with new technology investments against alternative policies employing existing technology. For example, we can compare the ESTS configuration as it would be used under the current system with the current configurations as they would perform under a consolidated repair policy.

Furthermore, a relaxing of policy assumptions may make different technologies viable. The ESTS program provides just such an example. Because it was required to deploy, it had to be designed to operate under harsh operating conditions. This requirement precluded consideration of less rugged commercial off-the-shelf (COTS) solutions.[14] Consolidated FSL options would have allowed COTS and other less rugged alternatives to be considered as well. If the analysis does show alternative ways of reaching the same goal, it should stimulate this type of thinking in future evaluations of new technology investments.

Our results and conclusions refer to two configurations of testers. The first uses the current inventory (either the F-15C or the F-15E configuration, as appropriate) and the second uses the ESTS for all F-15s.

---

[14]The Israelis use a COTS tester for the F-15I (the Israeli version of the F-15): the Electronic Branch Test Set (EBTS). It is not deployable but has much higher reliability than the current U.S. Air Force F-15 AIS testers. While it is not compatible with many of the LRUs that the Air Force will test with ESTS, by not strictly requiring a deployable tester at the start of replacement tester program, the Air Force may have further investigated how to make the EBTS compatible with F-15Cs and F-15Es. This information is based on interviews conducted in September 1998 with Master Sergeant Tim Flohrschutz, ACC Systems Office, F-15 Branch, and Major James Young, ESTS Program Manager.

# Chapter Three

# RESOURCE REQUIREMENTS DETERMINATION MODELS

## THE ACS EVALUATION GENERAL ANALYTIC APPROACH

The analysis and modeling efforts described herein followed the employment-driven modeling framework described in Chapter One. This chapter describes the methods that were employed to complete the first two steps, which together produce the logistics resource requirements to support operational plans. The third step—evaluating how the resulting resource requirements affect the goals—will be discussed in Chapter Five.

In this chapter, we first describe the mission requirements determination methodology used in the framework (Figure 3.1, left). We then present the models used in step two of the general approach (Figure 3.1, center), which we accomplish through a two-step process. First, a model converts the operational requirements into the demand for avionics components—the general support requirements. Second, a set of resource-specific models convert the general support requirements into individual resource requirements for maintenance equipment, maintenance personnel, transportation resources, spare parts, and infrastructure changes given a support structure design. Emphasis is placed here on models that either were newly developed or required substantial change or parameter modification.

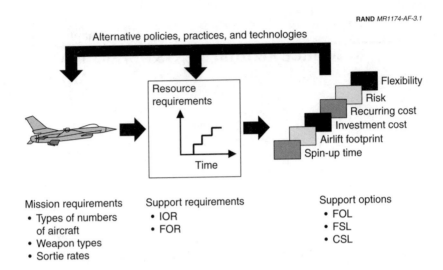

RAND *MR1174-AF-3.1*

Figure 3.1—A Standard Approach for Evaluating ACS Systems

## MISSION REQUIREMENTS DETERMINATION

To determine mission requirements, we relied on existing force employment scenarios from RAND MTW modeling, actual Air Force AEF packages, and Air Force war planning. In comparing MTW and small-scale AEF requirements, we found that the greatest single-FOL, regional, and Air Force-wide operating intensities all occurred in the two simultaneous MTW scenarios (the number of aircraft flying at one base, in a region, and across the Air Force and their operating tempo is the greatest in the two-MTW scenario). Since the support structure must satisfy the maximum mission requirements (expressed in terms of LRUs removed from aircraft), this two-MTW operating tempo dominance dictated the use of the two-simultaneous-MTW scenario to determine the support asset requirements of the F-15 avionics support system. The other mission types play a larger role in guiding how the assets should be employed.

To keep the report unclassified and thus more accessible, we elected to employ a notional RAND-generated two-MTW scenario, albeit one similar in operating tempo and deployment timing to U.S. war plans.

Actual Air Force spare-parts war-planning factors were used to determine the flying profiles.

## SUPPORT REQUIREMENTS

Existing models could not accommodate the translation of mission requirements into support requirements across the range of support structures considered in this analysis, particularly for maintenance equipment, personnel, and transportation requirements. Our efforts thus focused on developing a model that allows for the determination of tester, personnel, and transportation requirements for a given level of mission support. The result—the "maintenance shop requirements model"—provides the number of testers and people needed at a location as well as the daily volume of carcasses and spare parts that the transportation system must ship. When combined with the use of existing Air Force spare-parts models and simple infrastructure requirements models, this approach permits the generation of resource requirements for a given support structure.

The following sections describe the maintenance shop requirements model and go on to delineate how existing Air Force spare-parts models were used in the analysis, how output from the maintenance shop requirements model was used to generate transportation requirements, and how necessary infrastructure changes were estimated. The chapter entitled "Options Analysis" describes how the support requirements produced by these models translate to EAF goals through both quantitative metrics and qualitative assessments.

## MAINTENANCE SHOP REQUIREMENTS MODEL

The maintenance shop requirements model (Figure 3.2) has three components.[1] The first determines the maximum sustained demand the shop must handle (i.e., how many RRR LRUs must be replaced on aircraft in a given time period). The second ascertains the available

---

[1]For a complete description of the analysis method, see Appendix A.

RAND *MR1174-AF-3.2*

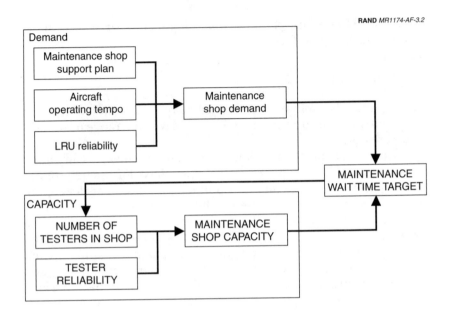

**Figure 3.2—Basic Structure of the Tester Requirements Model**

capacity in the shop (i.e., how many RRR LRUs it can produce in a given time period). The third yields the number of test stations that will be needed to provide sufficient shop capacity to reach the target maintenance wait time given the expected level of demand. The maintenance wait time is then combined with the actual repair time, the on-base distribution and carcass evacuation time, and, if appropriate, the FSL/CSL-to-FOL or home-base distribution and carcass evacuation time to produce the total resupply pipeline length for RRR LRUs successfully repaired by ILM activities.

In this model, demand for each test station type equals the number of test station hours required per day to repair all LRUs tested on a given type of test station. Daily demand is derived from the daily arrival rate at the test station and from the test station's usage duration per arrival for each LRU tested. Arrivals are determined from a combination of the number of aircraft to be supported, the operating tempo, and LRU removal rates per flying hour and sortie.

Capacity is a function of the number of test station hours available per day, which is determined by the number of test stations of a given type, tester uptime, and shop work hours per day. The following sections describe the data that the model requires and how those data are used to estimate demand and capacity.

## Demand on a Test Station Type

Figure 3.3 displays the data required and methods used to estimate test station demand. The flow inside the large box (top) shows the determination of demand for one type of LRU. The calculation starts with the daily demand rate from the Air Force's supply history—which, depending on the operating tempo, uses either the peacetime demand rate estimated in the Air Force supply history database (D041) or the decelerated wartime demand rate using wartime planning factors.[2]

The model uses supply system data to split the demand stream into AIS repairable and not-repairable-this-station (NRTS) components. However, there is additional demand not captured in the supply data—BCS removals. A removal is termed BCS when a problem cannot be found in an LRU and the part is returned to supply with no action taken (these are not recorded as removals in D041). Through manual data collection efforts by personnel at several bases, we were able to determine the BCS demand rate relative to repair rates. Using this relative rate with the D041 repair rates then produced a BCS rate. Additional demand on test stations also results from repairs that had to await parts (AWP), because the LRU must be put back on the tester when such parts arrive. This rate was estimated in a manner similar to the BCS rate.

The three rates—BCS, NRTS, and AIS repair (adjusted upward for AWP)—are then multiplied by the estimated average duration on the test station for each demand type. For BCS demands, the average

---

[2]Previous research indicates that LRU failure rates do not have a linear relationship to flying hours. In order to convert peacetime failure rates into projected wartime demands, the Air Force determined deceleration factors to convert longer wartime sorties into their equivalent peacetime durations in terms of LRU failure rates. The deceleration factors are specific to flying-hour scenarios. See Slay and Sherbrooke (1997).

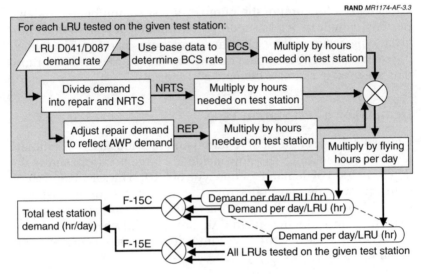

RAND *MR1174-AF-3.3*

D087 = Weapon system management information system

Figure 3.3—Test Station Demand Calculations

test station duration is estimated simply as BCS run time;[3] for the other two streams, it is estimated as twice BCS run time.[4] This yields the daily demand on the test station per flying hour for the LRU. Multiplying by the flying hours, decelerated to account for higher sortie rates for wartime scenarios, produces the average daily demand placed on the test station by the LRU in the operating scenario. This procedure is repeated for all F-15C and F-15E LRUs tested at each station. Summing the daily demand for all LRUs tested at the station then produces the total daily demand on that station.

---

[3] BCS run time was estimated through Air Force run-time standards and times provided by Seymour Johnson, Lakenheath, and Spangdahlem.

[4] For repairs, this time includes running through the test until the tester indicates a problem, making the repair, and then running through the complete test to verify that the LRU is serviceable. For NRTS events, a significant amount of time is often spent trying to make the repair before declaring the LRU NRTS. Personnel at several shops considered the estimates of twice the BCS time to be a reasonable assumption. Actual test station time for repairs was not available.

## Test Station Capacity

Test station daily capacity can be defined as the number of test station hours available per day for a given test station type. In the model, the uptime rates for a given test station type are constant regardless of the number of collocated testers. AIS testers can be capable of testing some LRUs while being inoperative for others, in which case they are termed "partially mission capable." To determine effective capacity, we compute an LRU weighted-average uptime for each tester type based on the daily demand and partial mission capability (PMC) rates for each LRU tested.

Detailed tracking performed prior to this analysis at two bases—Seymour Johnson and Lakenheath—revealed actual tester uptime by LRU for each test station. The weighted average PMC rate of the LRUs for the testers at these two bases serves as the average uptime in the model. Depending on the tester type, these bases have one, two, or three testers. Therefore, the uptime represents the expected uptime by tester for one, two, or three collocated strings (corresponding to the number of each tester type at these two bases) with the peacetime level of cannibalization practiced by these two shops.[5] Generally, the more testers a shop has, the more opportunity there would be to cannibalize one tester to keep the other testers operational. Since only one data point (a data point is a combination of tester uptime and the number of collocated testers) was available, however, for most testers we could not confidently estimate the uptime effect of cannibalization produced by varying the number of collocated testers. We therefore elected to use these calculated uptime rates regardless of the number of testers modeled in a shop.

Higher numbers of collocated testers would probably offer greater cannibalization benefit, so the model may slightly overestimate the number of necessary testers for support plans that consolidate maintenance resources. Similarly, if a squadron had to deploy independently with one string, the uptime would probably be lower than that used in the model because some level of cannibalization is probably occurring when all of the squadrons are home at Seymour Johnson or Lakenheath.

---

[5]Cannibalization is the act of using a part from a tester that is already down to keep another operational.

## Tester Quantity Calculation

The model minimizes the number of test stations subject to a maximum expected maintenance shop average wait-time constraint. In practice, this works by converting the maximum desired average wait time to a capacity utilization maximum that increases with the number of testers. As described in greater detail later, the more collocated testers, the greater the acceptable capacity utilization; the model simply adds testers until capacity utilization is less than or equal to the target, which also means that the expected average queue time is less than or equal to the target.

Current RSP calculations assume an average on-base LRU repair time of approximately four days. To be consistent with these assumptions, the capacity utilization in the maintenance shop requirements model is designed to produce an average maintenance queue time of 72 hours. This allows one shift for an average "repair action" (the average demand-weighted BCS time is 3.7 hours) and one shift for the combination of moving the LRU from the flightline to the AIS and moving the repaired LRU to base supply.

## PERSONNEL REQUIREMENTS MODELING

The method used to determine the personnel requirements of each structure builds on the maintenance shop requirements model, converting shop LRU demand into the number of test stations of each type needed to yield the required shop capacity. According to current Air Force practice, each tester used requires a team of two technicians. Using this as a standard and assuming that the shop must have sufficient personnel to operate all testers of a given type simultaneously when all are operational, we set direct-labor manning at two people per test station per shift.

## TRANSPORTATION MODELING APPROACH

Consolidated structures require transportation of carcasses and repaired LRUs between operating locations and repair activities. Currently, replacements of NRTS RRR LRUs and RR LRUs create shipping requirements. In the consolidated options, these requirements will not change, because we assume the same NRTS rate and set of

RR LRUs. However, any LRU that is now sent from the flightline to ILM for repair would instead be transported to an FSL or to a CSL. This includes not only LRUs requiring repair but also those later found to be serviceable upon testing (BCS). To simplify modeling, we estimate the transportation cost of the incremental shipping requirements introduced by consolidated systems rather than the total cost of each of the systems. This is the marginal transportation cost of going from a decentralized to a consolidated structure.

As an intermediate step, the maintenance shop requirements model estimates the average daily demand for each location for each LRU. We use the expected daily on-base repair volume (including those determined to be serviceable) for each LRU for each location to develop the incremental transportation requirements for each structure. This is based on the daily volume of LRUs currently being repaired on base that would have to be evacuated, repaired at a central site, and returned to base. Commercial express-service rates are used to estimate recurring peacetime transportation costs, based on the demand-weighted average LRU weight and the cube of these LRUs.

## SPARE-PARTS MODELING APPROACH

We used existing Air Force methods to compute required spare-parts levels, which are based on a model called Dyna-Metric, to determine the spare-parts requirements for each structure.[6] By adjusting parameters in standard Air Force models, these methods may in some cases allow for the modeling of policy and structure changes. In other cases, it is necessary to create new models, but where we do so, we employ the same assumptions and computation methods used in existing Air Force models. Discussed below are the models used for wartime and peacetime spare-parts requirements determination. Also examined are the differences in the structures that required changes either in model parameters or in the models themselves.

---

[6]See Pyles (1986).

## Wartime Readiness Spare Package Computations

The Air Force's Aircraft Sustainability Model (ASM), currently used to calculate RSPs, was used to determine the RSP requirements for each option.[7] We found it sufficiently flexible to calculate the RSP requirements for each option considered through changes in model parameters such as OST. Table 3.1 compares the RSP policies for the various structures used to set these parameters, with each row essentially corresponding to a parameter within the ASM. The left column describes the current policy used with the decentralized deployment option, and the middle column highlights the changes to existing RSP policy that must be made to render the RSPs appropriate for the decentralized-no-deployment option. The right column describes the RSP policy designed for compatibility with consolidated maintenance options. The next three sections describe changes that have been made in the RSP planning assumptions for each maintenance structure that required parameter changes in the ASM when calculating RSP requirements.

## RSP Policy for the Decentralized (Current) Structure

Under current policy, each squadron is designed to be self-sufficient for the first 30 days of a deployment, after which spare-parts resupply to the FOL commences. Thus, the RSP includes sufficient LRUs to cover the 30-day demand for RR LRUs, RRR LRUs that cannot be successfully repaired at the FOL, and SRUs to repair RRR LRUs. These calculations assume that the AIS will deploy and be operational by day three of flying operations at the FOL. Each squadron's RSP is based on the squadron's PAA, which is either 18 or 24 for all F-15 squadrons. The PAA combines with the peacetime removal rate, war mobilization plan (WMP) sortie rates, flying hours, and DSOs to project the wartime LRU removal demand rate, with DSOs expressed in fully mission-capable (FMC) aircraft availability on given days. The current WMP plans call for a DSO FMC floor of 63 percent on

---

[7]The ASM computes spare-parts requirements necessary to achieve specific aircraft availability objectives over a specified time horizon. It uses marginal analysis techniques to minimize the spares cost to achieve the target availabilities. For more information, see Slay et al. (1996).

Table 3.1

**RSP Policy Descriptions**

| Policy Factor | Decentralized (Current) | Decentralized No Deployment | Consolidated |
|---|---|---|---|
| Deployment | For RRR items, AIS deployed on day 0, operational day 3 | No AIS deployment, convert RRR items to RR | No AIS deployment, convert RRR items to RR |
| Supply | No resupply for 30 days | Resupply begins at day 1; 10 days from CONUS to OCONUS and 7 days from FSL to FOL | Resupply begins at day 1; 7-day resupply from FSLs; 10 days from CONUS |
| Stock consolidation | No centralized "risk" stocks | No centralized "risk" stocks | CSP computed to support operations of all RSPs |
| Unit size | 18 or 24 PAA RSP | 18 or 24 PAA RSP | 12 PAA RSP modules, with dependent 6/12 |
| Flying profile | War mobilization plan (WMP) sortie rates/durations | WMP sortie rates/durations | WMP sortie rates/durations |
| DSO | DSOs derived from WMP (63% FMC floor at day 30) | DSOs derived from WMP (63% FMC floor at day 30) | Min DSO of 75% at day 7 |

day 30. This means that RSPs are designed to have sufficient spare parts to keep squadrons' FMC rates at or above 63 percent on day 30 of flying operations, with equal or higher objectives on all other days of operation.

Since no resupply will occur for the first 30 days, the inventory position over that time consists only of what is on hand or in local repair (as opposed to the more typical case, where the inventory position includes inventory en route to the location and carcasses in the reverse pipeline). Also, since the POS inventory position at the time of deployment could potentially consist, in the extreme case, entirely of inventory due in from a depot, the Air Force has decided not to include any portion of the POS inventory level as a portion of RSP levels. Therefore, the on-hand inventory in the RSP must be equal to the full inventory requirement calculated in the RSP determination process.

## RSP Policy for the Decentralized-No-Deployment Structure

The dissociation of ILM from the FOL during wartime requires that several changes be made in RSP policy to render it consistent with maintenance policy. Inasmuch as there will not be any local repair, the RSP computations must treat all LRUs as RR, which in effect increases the depth requirement for all currently designated RRR LRUs (the operating stock requirement increases from a four-day local repair time to the time it takes an LRU to make the round trip from flightline removal to receipt at base supply). Removing ILM from the FOL also eliminates the need to have any SRUs in the RSP for repair at the FOL. Previous RSP modeling efforts found that making these changes and continuing to assume commencement of resupply on day 30 leads to excessive and unaffordable increases in spare-parts levels. Thus, changes in resupply policy would be necessary for this structure to be worth considering. In this analysis, we assume that supply begins at day one and takes ten days (under wartime assumptions) from CONUS to an FOL and seven days from an OCONUS location to an FOL in the same region (the baseline peacetime OST assumptions). While still requiring increased depth of the RRR LRUs, the increase becomes much less than if resupply were to begin at day 30.

## RSP Policy for Consolidated Structures

Current supply procedures and policies, while potentially optimal for the system for which they were designed, may not be the most appropriate for use with alternative maintenance structures. Since transitioning to a consolidated system would require changes to RSPs based merely on system design differences, we elected to exploit the required transition by also considering simultaneous changes aimed at making RSPs better oriented to AEF-type operations while retaining MTW capabilities. This analysis—which examined both the desirable RSP characteristics of the new AEF operational requirements and the system design characteristics of consolidated maintenance—resulted in the development of a new set of RSP policies.

As with the decentralized-no-deployment option, the major difference in spare-parts policy in this option is generated by the change in base-level (FOL) maintenance from RRR to RR. This changes the RSP from a mix of LRUs and SRUs to LRUs only. If the planned resupply time is met, the resupply time assumption does not affect operating performance but it does affect cost by increasing the pipeline length and hence the amount of inventory needed to cover demand over this time period. The longer the planned resupply time, the greater the number of parts needed in the pipeline.

The consolidation of support facilities set up for maintenance provides a new opportunity to make wartime spares stockage more efficient. The FSLs could also serve as consolidated stockage points for RRR LRUs by employing the CSPs described earlier, which would be buffer stocks located at regional repair facilities. They would be replenished by the local consolidated repair activity, providing for a relatively small pipeline requirement for CSP inventory at the FSLs. After preliminary modeling indicated that CSPs reduce the overall wartime spares requirements (the combination of CSPs and RSPs), we elected to model the consolidated options with CSPs. Without consolidated repair activities to replenish CSPs, they would have a lengthy pipeline from CONUS depots to primarily OCONUS regional supply FSLs, which would greatly increase the CSP requirement.

To better support small-scale AEFs, which often call for squadrons to provide only 6 or 12 aircraft packages, the new packages are designed in modular fashion. Base kits are designed to support 12 aircraft and supplemental kits are designed to support the residual 6 or 12 aircraft remaining in the squadron. These supplemental, or dependent, kits combine with the independent modular kit to provide sufficient support to cover a squadron for MTW operations. This makes it easy to split the RSP if only 6 or 12 aircraft deploy as part of an AEF package. The aircraft would have sufficient spares support while preserving flexibility to deploy the remainder of the squadron—also with sufficient spares support—at a later date to either the same or a different operating location.

With the small aircraft packages in AEFs, there is some question about whether the WMP DSOs are appropriate, because they would allow for a substantial percentage of an already-small force to be not mission capable (NMC). To enhance the effectiveness of these small force packages, we thus consider a higher DSO floor for RSP planning.

## Peacetime Operating Stock

For POS computations, we essentially used the same parameters for all maintenance structures; the only changes (see Table 3.2) were those reflecting the maintenance structures themselves (i.e., the location of ILM versus operating locations). By design, consolidated structures add a pipeline segment from bases to repair activities and create the potential for three echelons of supply (at bases, FSLs/CSL, and depots). Because the ASM can handle only two echelons of supply, we developed a model that uses the same underlying calculation methods as the ASM for computing pipeline stockage and safety stock quantities, but for three-echelon systems. The model employs the Air Force's standard base supply system (SBSS) formula to compute stockage levels.[8]

## DETERMINING INFRASTRUCTURE REQUIREMENT CHANGES

Infrastructure requirement changes consist of two components: facilities and permanent assignment locations of avionics ILM personnel. To determine new facility or facility expansion requirements, we assumed that planned FSL/CSL locations would need to add space only to accommodate increases in the number of test stations. After these increases were determined for each test station type, they were multiplied by the number of square feet per test station (per given type) specified in the Air Force specification for an AIS. Standard construction costs were then used to convert this square footage requirement into an investment cost. Administrative space was held constant.

---

[8]The underlying SBSS calculation is described in the *USAF Supply Manual*. The pipeline stockage quantity is based on the demand rate, OSTs, retrograde times, and repair times for a part and is essentially the amount of stock required to cover expected demands over the time it takes to receive a replacement for a part removed from an aircraft. The safety stock formula is $(3 \times p)^{1/2}$, where p is the pipeline quantity. Safety stock accounts for the variability in demand and replenishment time.

## Table 3.2

## POS Policies for Decentralized and Consolidated Structures

|  | Decentralized (Current) | Consolidated Options |
|---|---|---|
| Type of parts | SRUs for RRR LRU repair<br>RR and RRR LRUs | No SRUs<br>RR and RRR LRUs |
| Baseline OST | 7 days CONUS<br>10 days OCONUS | 7 days from FOLs to<br>CSL/FSL |
|  |  | 7 days from CSL/CONUS<br>FSL to depot |
|  |  | 10 days from OCONUS<br>FSL to depot |
| Model used for safety-level formula | ASM | SBSS |
| Stock locations | Bases, depots | Bases, FSLs/CSLs, depots |
| Removal rate data | D041 | D041 |
| Repair times | D041 base repair time | D041 base repair time |

In the event that the Air Force elected to provide entirely new buildings designed for increased AIS size, the facility cost would approximately triple. Conversely, if existing buildings could be used to accommodate the AIS size increase, facility costs would decrease.

In each of the consolidated structures, some personnel would have to move from bases losing their AIS to FSL/CSLs. This number was determined by comparing the personnel requirements at each FSL/CSL to current personnel levels. Multiplying this by the standard Air Force rate for the cost of moving people in a unit move produces the total permanent change-of-station cost.

# Chapter Four

# RESOURCE REQUIREMENTS

## OPERATIONAL EMPLOYMENT REQUIREMENTS

The first step in the employment-driven analysis approach is to assess the mission requirements analysis to determine the peacetime and wartime beddowns and flying requirements for F-15s worldwide. The left side of Table 4.1 lists the notional 2004 peacetime F-15 beddown used in the analysis, which shows primary aircraft authorizations (PAAs) for the active component, including combat-coded, test, and training aircraft. The right side of the table shows the notional deployment requirements to simultaneously support two MTWs used in this study. The shading indicates that there are too few F-15Cs to accommodate two MTWs simultaneously. It is assumed that there is enough time between the start dates of the two MTWs to allow F-15Cs to swing between theaters after establishing air superiority in the first MTW location. The peacetime flying hours used in this study are based on current peacetime flying-hour rates, and the wartime operating profiles are based on current Air Force wartime sustainment planning.[1]

---

[1]Combined with the deceleration factors discussed in Chapter Three, the wartime profiles effectively increase the daily avionics LRU demand on the system by about two and one-half times per aircraft. The exact flying hour and sortie rate profiles are not shown owing to their sensitive nature.

**Table 4.1**

**Notional F-15 Beddown and MTW Requirements**

| F-15E | Combat | Peacetime Noncombat | Total | MTW A | B |
|-------|--------|---------------------|-------|-------|---|
| Location 1 | 48 | 36 | 84 | 24 | 24 |
| Location 2 |  | 3 | 3 |  |  |
| Location 3 | 18 |  | 18 |  | 18 |
| Location 4 |  | 10 | 10 |  |  |
| Location 5 |  | 2 | 2 |  |  |
| Location 6 | 18 |  | 18 | 18 |  |
| Location 7 | 48 |  | 48 | 24 | 24 |
| Total | 132 | 51 | 183 | 66 | 66 |
| **F-15C** | **Combat** | **Noncombat** | **Total** | **A** | **B** |
| Location 1 | 18 |  | 18 | 18 |  |
| Location 2 | 48 | 10 | 58 | 24[a] | 48[a] |
| Location 3 | 66 |  | 66 | 18[a] | 66[a] |
| Location 4 |  | 16 | 16 |  |  |
| Location 5 |  | 76 | 76 |  |  |
| Location 6 |  | 3 | 3 |  |  |
| Location 7 | 42 |  | 42 | 36 | 6 |
| Location 8 | 24 |  | 24 |  | 24 |
| Location 9 | 48 |  | 48 | 48 |  |
| Total | 246 | 105 | 351 | 144 | 144 |
| **Total** | **378** | **156** | **534** | **210** | **210** |

[a]Includes swing forces that must shift from one theater to another.

## MAINTENANCE REQUIREMENTS

Entering the peacetime and wartime demands into the maintenance requirements model generates the number of testers required at each location during peacetime and wartime. Summing across all locations for each structure for the two-MTW scenario then provides the maximum number of testers required at any given time for each structure. The top graph in Figure 4.1 shows the results using the current tester configurations. Full decentralization, for example, requires 157 test stations. The bottom graph shows what the ESTS requirements would be if the ESTS achieved its planned 95 percent availability.[2] In this case, full decentralization would require 127 test

_____

[2]See Armstrong (undated).

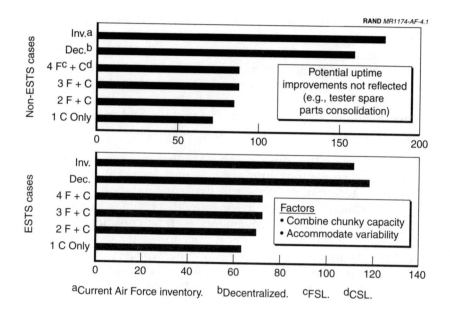

Figure 4.1—Tester Requirements for Each Structure

stations, including 42 ESTS units, that must be combined with the requirements for the antenna system (either EARTS or Antenna A and Antenna B), TISS, and EAU. The results are based on the assumption that capacity will not be provided during combat operations to test LRUs designated as two-level items, although units could use excess peacetime capacity to test these two-level LRUs (as is the practice now).

The inventory line (Inv., first row) in the top graph of Figure 4.1 indicates the number of testers in the Air Force inventory currently supporting active forces at air force bases (excluding depots, Air Force Reserve, and Air Force National Guard testers). For the ESTS graph, the top row reflects the planned inventory.

## REDUCTION OF TESTER REQUIREMENTS

The graphs in Figure 4.1 show the effect of consolidation on tester requirements. The current decentralized structure, for example, requires 31 TISS testers, but this requirement drops by about one-third for all levels of consolidation. Because capacity comes in steps equivalent to one tester, excess capacity at each location declines as consolidation increases. In addition, the greater the number of collocated testers, the greater the maximum capacity utilization that will lead to a given maintenance wait time, because consolidated testers accommodate capacity (tester uptime) and demand (failure rates and repair time) variability more effectively.

The estimates for the consolidated ILM cases are likely to be conservative because we took into account only stepwise capacity and variability accommodation[3]—the two consolidation factors that improve "effective capacity" in a manner that could be quantitatively estimated. Potential improvements such as consolidating tester spare parts and increased opportunity for tester cannibalization would likely improve overall tester uptime, which might further reduce tester requirements in the consolidated cases. In this analysis, we held average tester uptime for each LRU constant for all maintenance options.

## Consolidation Effect of Discrete Capacity Levels

Consolidation effects result from "chunky" capacity and variability accommodation. Figure 4.2 illustrates the consolidation effect resulting from the stepwise nature of capacity in this type of process. Capacity does not increase smoothly but rather in steps. Each step in the figure represents the addition of a tester. The horizontal line just above the 24-hour tick mark represents a notional demand on one tester type created by a squadron. In this case, demand is just above the capacity of one tester, so two testers would be needed to support a base with one squadron; otherwise, the maintenance backlog would continuously grow. For decentralized maintenance in support of four noncollocated squadrons, eight testers operating at an average of 55 percent capacity would be required, meaning that almost half of the capacity at each location would be excess. Consolidation takes advantage of this distributed excess capacity by combining it into usable chunks. As a result, the pooled demand of

---

[3]We call capacity stepwise or chunky because capacity can be increased or decreased only by multiples of the capacity of one tester.

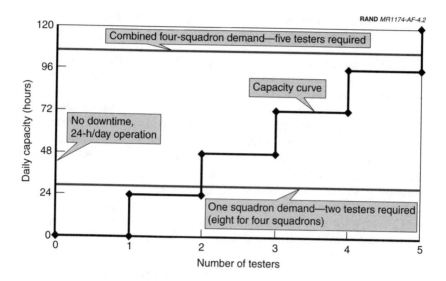

Figure 4.2—The Step Function Effect of Tester Capacity

four squadrons requires only five collocated testers in this example, reducing the tester requirement by 38 percent and boosting capacity utilization to 85 percent.

## Consolidation Effect of Load Smoothing

Consider a system having five sets of customers with equal demand rates and five servers with equal capability. In system one, the sets of customers and servers form five subsystems consisting of five one-server queues. In system two, they form one system with one queue. In the presence of service time or demand variability, system two will have a shorter average wait time. The number of system-one servers would thus need to be increased to achieve the same average queue time as in system two. This is because if one server in system one is backlogged and one is idle, the idle server cannot help out the other. System one is akin to a decentralized maintenance system in which each squadron is supported by one set of testers. By contrast, system two represents consolidation. Since each structure was assumed to have the same maintenance queue time, the consolidated structures need fewer testers to satisfy the same aggregate system demand as a decentralized structure.

The ASM spares calculations assume that base repair times would remain constant at current levels, averaging about four days across all LRUs. In the interests of producing modeling consistency, the capacity utilization factors were set for an average maintenance queue time of 72 hours, with the remaining repair time split between actual repair time and on-base movement. Given these parameters and assuming the same level of demand, a consolidated system needs about 10 percent fewer testers than a decentralized system. This consolidation effect is modest because 72 hours is a relatively long queue time compared to the service time. Reducing the queue time to 12 hours, for example, would give a consolidated system close to a 50 percent advantage in tester reduction over a decentralized system.

## SHORTAGES OF CERTAIN TESTERS IN TODAY'S SYSTEM

The model's results also provide an opportunity to evaluate the ability of the current structure to meet its requirements. Our analysis suggests that the Air Force would need six additional TISS testers to meet the demands of two MTWs. For all other current testers, the inventory is sufficient. The model also shows, however, that the proposed purchase of 39 ESTS may not suffice to meet current maintenance objectives, even if the ESTS achieves 95 percent availability.[4] The model estimates that the Air Force would need to purchase three more ESTS testers to meet operational objectives.

### Excess Testers Permit MTW Tester Capacity During Peacetime at All FSLs

The tester requirements displayed in Figure 4.2 represent requirements across the Air Force during two simultaneous MTWs (the maximum total Air Force requirement). However, the sum of the maximum required at any one time at each location in the consoli-

---

[4]Sensitivity analysis of ESTS uptime reveals that in the decentralized case, the number required would not increase until availability dropped below 83 percent. This unexpected result occurs because at 95 percent availability the stepwise nature of capacity gives significant excess capacity to each distributed location.

dated structures (peacetime for a CSL with FSLs and during an MTW for FSLs) is always less than the current number of testers that the Air Force has acquired or plans to acquire under the decentralized structure. Consider, for example, the support structure of two FSLs and one CSL. For an MTW, each FSL needs eight TISS testers. The CSL needs three testers during two simultaneous MTWs and seven during peacetime. The sum of the maximum number of testers required at any one time at each of the three locations is therefore 23. Since the Air Force currently has an inventory of 25 TISS testers, it could permanently equip the CSL and both FSLs at their maximum projected requirements. This is possible for all testers in any of the consolidated structures. Relying on this "excess" tester inventory in the consolidated structures would eliminate the need to move any testers across all scenarios.

## PERSONNEL REQUIREMENTS

The model calibrates the personnel requirements for each structure using current Air Force manning standards. Given the current system, practices, and personnel levels, we estimate that two people would be needed for each shift at each test station. The number of test stations indicated by the maintenance shop requirements model thus determines the number of personnel required to support operational objectives for each structure.

Figure 4.3 displays the personnel requirements for both tester configurations under each support structure. Decentralized support personnel requirements are significantly greater under all tester configurations, even with the reduced personnel demands of an ESTS shop.

The thick horizontal line in Figure 4.3 indicates the current level of Air Force F-15 avionics AIS personnel. As with testers, for the two-MTW case under decentralization our analysis points to a personnel shortfall comprising almost 300 persons for the current AIS technology and about 50 persons using the ESTS technology. The current manpower level is close to, but still slightly below, what our model indicates is required for the ESTS configuration. The difference between the modeled ESTS configuration requirement and the number of personnel today results from the need for six additional TISS and three additional ESTS test stations beyond today's TISS and the

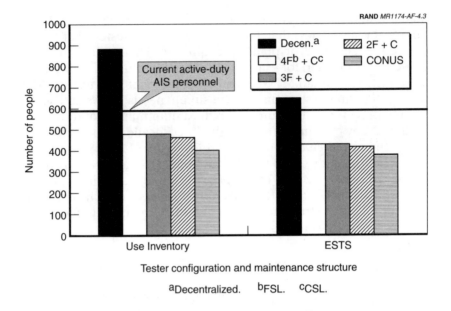

RAND *MR1174-AF-4.3*

Figure 4.3—Personnel Requirements

programmed ESTS inventory. In effect, the Air Force has already reduced personnel to the level necessary to operate the number of programmed ESTS test stations even though the "old" testers are still in use.

## SPARE-PARTS REQUIREMENTS

In the decentralized structures, spare parts must accommodate base repair of three-level LRUs, three-level LRUs declared NRTS, and two-level LRUs at depots. On-base stock includes the base-repair LRU operating requirements and safety stock based on those requirements, SRUs to support the replenishment time from the depot, and operating and safety stock to support LRUs sent to depots. Depot stock must include an operating and safety level of SRUs to repair LRUs based on SRU replenishment time, safety stocks of LRUs based on the depot repair time, and operating and safety stocks of LRUs to replace condemned LRUs. The only stock based on the depot-to-base pipeline time is the on-base stock supporting NRTS LRUs (not all two-level items are returned to depots for repair during peacetime).

In the consolidated structures, spare parts must accommodate FSL or CSL repair as well as depot and home-base requirements. Bases must still have LRU operating and safety stock, but the levels are based on the pipeline to consolidated repair centers for three-level items rather than on the on-base repair pipeline. The same pipeline between depots and bases is used for two-level items, and there is no longer a need for SRUs in RSPs or in POS at bases not collocated with a repair facility. FSLs and the CSL must have SRU operating and safety stock to support the SRU replenishment time from the depot, as well as LRU operating and safety stock to support the on-site repair time and NRTS LRUs. Depot stock in the consolidated structures would cover the same pipelines as those in the decentralized structure. In the consolidated structures, both base and FSL/CSL stocks rely on resupply times from FSLs/CSL and depots, respectively. Figure 4.4 illustrates the difference in spare-parts operating stock pipelines between having a collocated AIS and receiving support from a regional FSL in theater.

Reflecting these requirements, the spare-parts levels in Figure 4.5 were produced by the ASM- and SBSS-based models as appropriate. Counterintuitively, we see that as consolidation increases, spare parts requirements increase. This is attributable to the fact that as consolidation increases, the average pipeline length across the Air Force for POS increases as well because fewer aircraft are collocated with ILM. During peacetime in the two-FSL and one-CSL case, for example, the F-15s at Lakenheath and Kadena have local repair with a pipeline of just four days, but in the one-CSL-only case they must receive all support from CONUS with an effective pipeline length of 24 days with the base-case OST of ten days (ten days retrograde, four days of CSL repair time, and ten days of shipment time to the base) and even ten days with the "faster" OST of three days.

For RSPs, each of the FSL structures has the same inventory requirement because the only factor that changes across the options is wartime pipeline length. For all options, this is ten days from FOLs to CONUS locations and seven days from FOLs to FSLs. Therefore, the RSP requirement is the same for each of the three FSL options

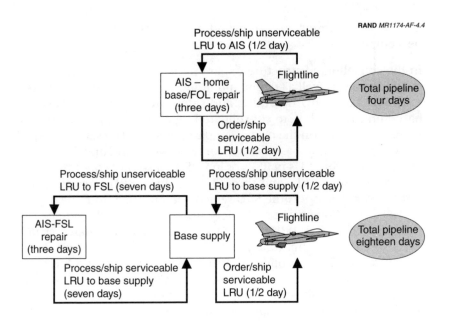

RAND *MR1174-AF-4.4*

**Figure 4.4—Comparison of Spare-Parts Operating Stock Pipelines Between
Having a Collocated AIS and Receiving Support from an FSL
in Theater Using the Base-Case OST of Seven Days**

and largest for the CSL-only and decentralized-no-deployment options. In consolidation, none of the aircraft will be collocated with
repair during wartime, so this factor, which strongly influences POS
requirements, plays no role in RSP computations.

Consolidated repair offers inventory reduction opportunities
through potential reductions in repair time and through reductions
in safety stock created by pooling demand. With long repair times at
bases, substantial operating and safety stock requirements would
offer a large potential for improvement. In terms of the operating
stock requirement, if consolidation can reduce repair time by more
than the increase in distribution time from the repair location to the
operating location, then consolidation reduces inventory requirements. Reductions in safety stock both from repair time reductions
and from pooling of demand would combine with changes in the

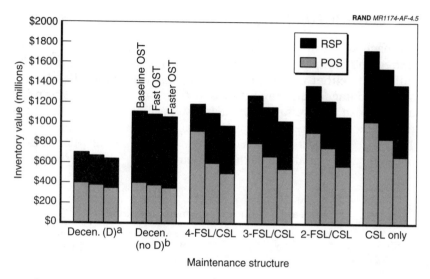

aDecentralized—deployment option.    bDecentralized—no-deployment option.

**Figure 4.5—Total Inventory by Structure**

operating stock requirement to produce the total change. The current short Air Force local repair times of about four days do not offer much room for improvement in operating stock and result in fairly low safety stock levels. Even reducing consolidated repair time to one day would result in an increased pipeline length unless distribution times of less than two days could be achieved.

## PEACETIME TRANSPORTATION REQUIREMENTS

During peacetime there are about 180 LRU demands (including BCS) per day across the system. Depending on the structure, some portion of these will require evacuation to repair facilities and transportation of serviceable LRUs to home operating bases. Summing across the demands requiring transportation at all locations in each structure produces the daily transportation requirement.

## INFRASTRUCTURE REQUIREMENTS

For each structure, we determined the number of tester types required and compared it to the on-hand total at each location. The difference, either positive or negative, was then multiplied by the square-footage requirement for that test station type.[5] Summing across these requirements for each tester type at each location produces the facility requirements for each structure, as indicated in Table 4.2. Any existing unused space at the current facilities at each of these bases would reduce these requirements.

Table 4.2

### Increase in Square-Footage Requirements

| Current Testers | 4-FSL/CSL | 3-FSL/CSL | 2-FSL/CSL | CSL only |
|---|---|---|---|---|
| Kadena | 11,615 | 11,615 | 11,615 | — |
| Lakenheath | — | — | 8675 | — |
| SWA | 24,814 | 24,814 | — | — |
| CSL | 4609 | 7157 | 7157 | 32,612 |
| ESTS | | | | |
| Kadena | 5368 | 5368 | 5368 | — |
| Lakenheath | — | — | 1544 | — |
| SWA | 23,489 | 23,489 | — | — |
| CSL | — | 219 | 219 | 22,422 |

Similarly, we determined personnel movement requirements. The number of people currently assigned at each location was compared to the consolidated requirement for each option. The requirements were then summed across locations to yield the total movement requirements for each structure displayed in Table 4.3.

---

[5]Square-footage requirements for each test station type were determined by dividing the total shop space dedicated to each tester type by the number of test stations in the general AIS building specifications found in "Basic Item 217-712, Avionics Shop (ADDM-D)," McDonnell Aircraft Company, MDC A9246, Rev. B.

### Table 4.3

### Required Personnel Moves for
### Implementation

| Configuration | Current Testers | ESTS |
|---|---|---|
| 4 FSLs + CSL | 245 | 185 |
| 3 FSLs + CSL | 284 | 224 |
| 2 FSLs + CSL | 268 | 208 |
| CONUS | 325 | 295 |

# Chapter Five

# OPTIONS ANALYSIS

This chapter discusses the costs and deployment footprint requirements for each structure as well the effects of these requirements on risk, flexibility, personnel turbulence, and spin-up time. Although a disproportionate amount of this chapter describes the costs resulting from these resource requirements, it is the total cost that serves as a feasibility check of sorts on each structure. In today's budget environment, it is highly unlikely that the Air Force would elect to increase support costs even if it were capable of doing so. Thus, it is critical to determine if alternative support systems are even feasible from a cost perspective.

## MINIMIZE COST

The five costs analyzed here are for testers, personnel, transportation, spare parts, and infrastructure changes.

### Tester Costs

As a result of existing or programmed inventories, which are treated as sunk costs, tester procurement costs are minimal. Figure 5.1 summarizes the tester costs for both tester configurations, using current inventory and the ESTS, under each support structure option. Procurement costs for the existing testers are incurred only under continued use of decentralized support, since there is only an insufficient number of testers in current inventory to meet MTW objectives with this system. For both tester configurations under decentralized support, there is a $12 million cost associated with increasing the number of TISS testers.

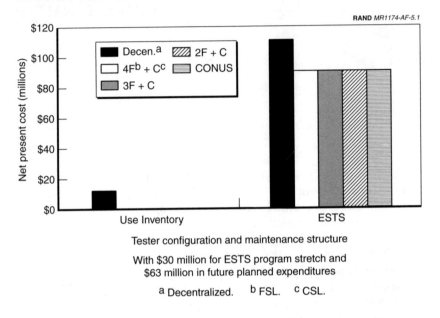

Figure 5.1—Tester Costs by Structure and Tester Configuration

Although parts obsolescence and the potential need for overhauls are likely to introduce some cost risk in using the current testers, we assume this cost to equal that of the current programmed tester maintenance budget and therefore do not consider it in this analysis.

ESTS configuration costs include $63 million in program funds still pending when this analysis was conducted (FY99). Assuming that these funds could have been shifted helps illustrate the tradeoffs that can be made between technology and policy. These costs also include a then-projected program-stretch cost of about $30 million. This stretch is intended to address the need for modifications to eliminate tester deficiencies uncovered during operational test and evaluation. The decentralized ESTS case includes the costs of procuring three ESTS units in response to the projected shortage discussed earlier. With the need for TISS testers in the decentralized structures, this brings the ESTS configuration tester cost to about

$115 million, as opposed to $93 million in the consolidated structures.

## Personnel Costs

Personnel costs are based on an average cost of approximately $42,000 per person per year, weighted for the desired mix of skill levels and grades of authorized personnel. These costs reflect fully burdened Air Force personnel costs.[1]  Figure 5.2 compares the present value of personnel costs over an eight-year period discounted at a 5.5 percent annual rate (eight years is estimated to be the economic life of test equipment).  Since personnel cost is a recurring cost that can be adjusted over time, it varies much more between structures than do tester costs.

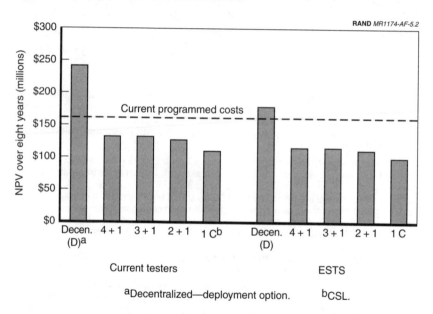

**Figure 5.2—Eight-Year Net Present Value of Personnel Costs**

---

[1]By "fully burdened costs" we mean the sum of all costs associated with having a person in the Air Force, to include pay, benefits, accrued pension expense, infrastructure costs (e.g., base housing), and support costs.  See "Application of Military Standard Composite Rate Acceleration Factors for Fiscal Year 1998."

## Spare-Parts Costs

To determine spare-parts costs, we start with the total requirements for each part, the sum of which was displayed earlier. By comparing these requirements to the current Air Force inventory, programmed buys, and programmed repairs, we then determine the changes in procurement and repair requirements necessary for each LRU and SRU for each structure.[2]  In cases where there is long supply of un-serviceable parts, incremental requirements are met first through repair and then through procurement once the new requirements consume the excess unserviceable stock. Adding the change in requirements for all LRUs and SRUs then produces the total additive outlays for buy and repair displayed in Figure 5.3.[3]  These additive outlays represent one-time infusions of money that would have to be directly appropriated to AFMC through the Air Force Working Capital Fund (AFWCF) to increase authorized serviceable inventory levels through a combination of one-time increases to repair programs and new procurements.[4]  Most of the additive outlays are for buys that must

---

[2]The central secondary item specification (CSIS) was used to determine the actual inventory on hand, due in from procurement, or in long supply for each item. Many F-15 avionics spare parts are in long supply Air Force-wide, but much of the "excess" inventory is in storage at depots in unserviceable condition awaiting induction into repair as demands occur.

[3]Air Force programmed buy-and-repair costs for each part were used to determine costs. Programmed repair costs include necessary piece-part costs to repair the component as well as "overhead" and labor cost allocation. Since in the short run overhead and labor costs are primarily fixed costs, actual repair-program changes may not cause changes in actual costs equal to the estimated costs using the programmed repair costs. Because programmed repair costs are orders of magnitude lower than programmed buy costs, however, almost all of the net outlays from spare costs are produced by net procurement changes. Therefore, the possible programmed repair cost "misrepresentation" of actual marginal repair cost has an insignificant effect on the comparison of options in this analysis.

[4]Spare-parts repairs and procurements are typically paid for through the AFWCF. This revolving stock fund operates on a full cost recovery basis that is intended to allow the AFWCF to break even. (By law it cannot make a profit and if it does so in one year, the profit is returned to customers in the next year through a reduced overhead charge. Similarly, if there is a loss in a year, it must be made up through an increase in the overhead charge the following year. Alternatively, a one-time appropriation is sometimes made to make up losses in order to avoid increased unit prices, which may drive customers to look for alternative sources of supply [e.g., local repair of items when they should come from a depot].) When a unit draws an item from stock, it pays the AFWCF an amount equal to the marginal cost of the item plus an overhead markup. This money should then be used to repair the carcass (or purchase a new

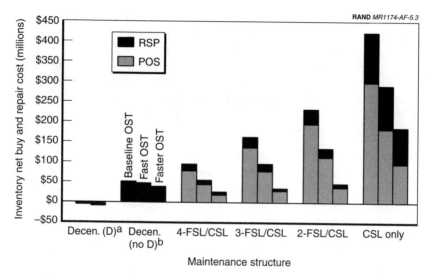

Figure 5.3—Additive Inventory Repair and Procurement Outlays

occur after unserviceable LRUs are exhausted, because programmed repair costs are extremely low relative to the purchase price of many of these LRUs.

---

part if the carcass gets condemned) and replenish the inventory account from which the serviceable part was drawn. Moving to one of the alternative systems would require a one-time increase to new steady-state inventory levels. This represents a one-time unfunded demand on the AFWCF, because the repair and procurement requirements come not from customer demands that produce revenue, but rather to increase stock positions. Thus, to achieve the initial required position, the AFWCF would have to receive sufficient funds through direct appropriation to complete the required repairs and procurements.

This analysis also assumes that the RSP and POS requirements for the current system are fully funded. For a variety of reasons (all beyond the scope of this report but currently being worked intensively by AFMC and the Air Staff), there have been times when the Air Force Working Capital Fund (AFWCF) has not fully recovered its costs and the repairs and procurements required to maintain serviceable inventory levels have not been funded. As any such shortfalls should be funded regardless of the alternative selected, they do not affect the cost comparison between the alternatives. And since these potential costs are unrelated to a potential change in F-15 avionics maintenance structure, we elected not to include them in the analysis.

When determining outlays, we first pool the RSP and POS requirements and then compare the total requirement against the spare
parts program. We then "allocate" the outlays to RSP and POS for
analysis purposes. If only the POS or the RSP has an increased requirement for an item, then all of the outlay is allocated against the
respective inventory category. If both require an increase in the requirement, we allocate the outlays in the relative proportions that
POS and RSP contribute to the total requirement.

As with the overall spare-parts requirements, required spare-parts
outlays increase with consolidation. The increase in spare-parts
costs from the five-location to the three-location structure is linear
and caused by a linearly increasing number of aircraft supported by
FSLs rather than their bases.[5] The large jump in spare-parts costs
from the two-FSL/one-CSL option to the one-CSL-only option reflects both an increase in the number of aircraft supported from a
consolidated center (rather than from their home base) during
peacetime and the shift in OCONUS resupply from theater FSLs to
the CSL, which increases the pipeline in each direction from seven to
ten days using the baseline OSTs.

As expected, reducing the peacetime OST dramatically decreases the
POS spares inventory requirements and consequently the additional
outlays for buy and repair in the consolidated options. At the faster
times, this leads to the five- to three-location structures having similar outlay requirements. Even at the faster times, however, the CSL-
only requirement is substantial.

---

[5]The reader will note from earlier discussions that only the total POS requirement
changes between the three FSL options (there is no change at all in the RSP requirement). However, the POS and RSP requirements "interact" when determining outlay
requirements. In the four-FSL option, for example, consider a case in which the requirement for both POS and RSP increases but the current program can satisfy the entire POS and RSP marginal requirements (through "excess"). Suppose, however, that
in the three-FSL option, a further increase in the POS requirement causes the "excess"
position to run out and there is now a need for additional outlays of money to increase
the program. We allocate the outlays proportionately according to the POS and RSP
inventory increases. So now the total RSP outlay in the three-FSL option is greater
than that in the four-FSL option even though the total amount of RSP inventory is the
same. In this case, by themselves there would be enough inventory to satisfy either
the POS or the RSP requirements. Together, there is a need to increase the buy-and-
repair program.

For the decentralized case, improving the OST does little to reduce planned outlays, because current inventory is a sunk cost and most LRU resupply comes from repair at either bases or depots rather than through procurement. Thus, the reduction in requirements leads primarily to a delay in repair requirements. This creates a small, one-time savings through "repair avoidance" (see Figure 5.3).

In the long term, however, reducing OST would probably yield an inventory cost benefit exceeding that quantified in this analysis. When force modernization introduces new parts, the initial stockage requirement and hence the initial procurement for these parts would be lower with a reduced OST. Changes in forecasted demand would have a smaller impact on required changes in inventory levels, whether up or down. This would potentially reduce inventory "churn," which often produces "excess." In short, anything that affects required inventory levels in the future would have a smaller effect, because inventory levels would be lower. However, this would apply to each of the structures.

## Transportation Cost

Figure 5.4 displays the peacetime cost differences between the consolidated structures and today's decentralized structure. In a decentralized maintenance system, unserviceable three-level items are repaired on base and do not need transportation. In a remove-and-replace system, all unserviceable items must be shipped from FOLs or bases to an FSL or CSL, and a serviceable part must be shipped back. As consolidation increases, peacetime transportation cost increases as well because fewer operating bases will have collocated repair facilities. Transportation costs reflect Federal Express rates and can thus be assumed to correspond to the fast-OST scenario, serving as an upper bound for the baseline OST cost. The Air Force, however, is more likely to reach the fast-OST assumptions by reducing the order, depot, and receipt-processing segment times rather than the actual transportation segment, since often this already relies on express delivery for critical requisitions.

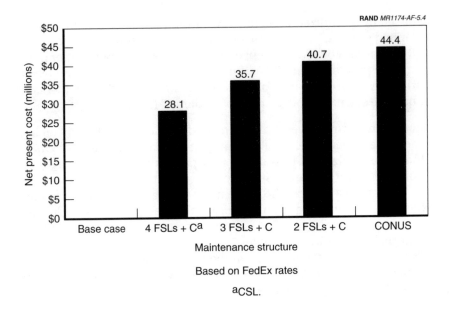

RAND *MR1174-AF-5.4*

**Figure 5.4—Eight-Year Net Present Value of Marginal Peacetime Transportation Costs**

## Infrastructure Change Costs

Figure 5.5 provides the infrastructure change costs by structure divided into facility and personnel/unit transfer costs. To determine facility costs, we simply multiplied the square-footage requirements by a standard cost per square foot for aircraft hangar construction and setup—$184 per square foot.[6] The square-footage requirements were based on the amount of space needed for test station requirements beyond the existing level at each location. Any existing administrative, storage, and other nondirect testing space was assumed to be sufficient. To determine the costs to move people from current duty assignments to the FSLs/CSL, we multiplied the number of

---

[6]The cost per square foot came from a table of historical averages provided on the Building Evaluations website. The cost per square foot for hangar-aircraft maintenance is among the highest of all building types. See "Hangar-Aircraft Maintenance" (2000).

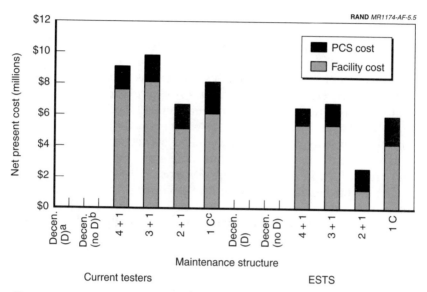

RAND *MR1174-AF-5.5*

aDecentralized—deployment option.    bDecentralized—no-deployment option.    cCSL.

**Figure 5.5—Infrastructure Change Costs by Structure**

people who have to move by the standard FY99 cost per person for unit moves—$11,517 for officers and $5,881 for enlisted personnel.[7]

Facility costs are a function of the number of maintenance locations and the amount of expansion that each needs. The primary reason that the four- and three-FSL options are more expensive than the others is that they include a notional SWA FSL. The SWA FSL facility cost is much higher than the others because we assume the need to construct a complete facility rather than just an addition with space for an increased number of test stations as with the other locations. The ESTS configuration options are less expensive because adoption of the ESTS reduces the total number of test stations at each location.

---

[7]"Permanent Change of Station (PCS) Cost Per Move As of FY98/99 President's Budget," (1998).

## Total Cost

Summing the net present costs for equipment, personnel, spare parts, transportation, and infrastructure changes for each option produces the total costs for each structure, as shown in Figure 5.6. With baseline OSTs and the current tester configuration, the decentralized-deployment structure and the four-FSL option are the lowest in total cost. However, these two structures differ greatly in cost components, with recurring personnel costs accounting for most decentralized structure expenses and spare-parts investment comprising a higher proportion of the four-FSL cost. This reflects the general tradeoff between consolidation and decentralization—spare-parts investment versus annual personnel expenses. As consolidation increases, the spare-parts cost increases substantially while personnel

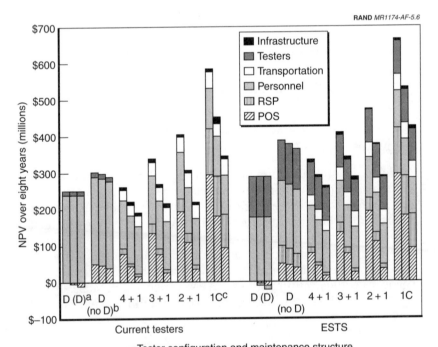

Figure 5.6—Total Cost by Structure, OST, and Tester Configuration

savings accumulate at a slower rate beyond the first jump to consolidation. As a result, the consolidation option with the least amount of consolidation, the four-FSL one-CSL option, has the lowest cost of the four.

For baseline OSTs with the ESTS configuration, decentralization carries a slightly lower cost than the four-FSL option because of lower ESTS personnel requirements compared to the current testers given the same demand and maintenance structure. Since the main cost benefit of consolidation is a reduction in personnel (and in testers if they have not been procured), introducing ESTS or other changes that reduce personnel decreases the recurring-cost advantage gained solely through consolidation.

Under fast peacetime OSTs, spare-parts requirements decline while other costs remain the same.[8] With the current testers, the four-FSL option has the lowest costs under fast OSTs. With the ESTS, using fast OSTs makes the four-FSL and decentralized-deployment structures approximately equal in cost. Using "faster" OSTs further cuts spare-parts costs while again holding other costs steady. With the faster OSTs and current testers, several consolidation options would cost significantly less than would the decentralized deployment option. Using the ESTS, these options would cost roughly the same as decentralized deployment.

In comparisons of different tester configurations, remaining ESTS program funds are critical. If such funds are recoverable, then the current testers are more cost-effective. If they are not recoverable, then the ESTS becomes more cost-effective because it reduces personnel requirements. In this case, the ESTS program cost would have to be treated as a sunk cost, leaving only the additional required tester cost in the decentralized case as the only marginal tester cost. Our evaluation of the current inventory configuration, however, assumes that no investment in the current testers is necessary (except for the addition of six TISS testers in the decentralized option), so any

---

[8]The reader should note from an earlier discussion that the change in peacetime OST affects only the absolute requirements for POS and has no effect on RSP absolute requirements. As discussed previously, however, the interaction between the two when determining marginal requirements results in the small changes in the marginal RSP requirements seen in Figure 5.5.

investment resulting from obsolescence would reduce the cost difference between the configurations. The ESTS configuration also carries cost risks based on uncertainty in operational requirements and program stretch costs.

## REDUCE PERSONNEL TURBULENCE

As the result of personnel retention problems, the Air Force has had difficulties in maintaining the required skill mix of personnel in some areas, such as F-15 avionics intermediate maintenance personnel. To solve this problem, the Air Force can either achieve better retention of its current personnel or it can find other sources of maintainers.

One of the goals of the EAF is to produce more predictable deployments for personnel and to ensure that people have periods of non-deployment stability toward the goal of improving retention. In this analysis, we take this one step further by focusing on reducing personnel deployment requirements to support operations as well as by examining where personnel must deploy—to FSLs or FOLs. The current structure, based on deployment of the AIS, produces high deployment requirements, often to hostile locations, whereas the de-centralized-no-deployment option completely eliminates deployment.

For the consolidated structures, different peacetime personnel basing options lead to varying levels of deployment requirements. Worldwide, the two-MTW personnel requirements are about 50 percent greater than peacetime requirements. For the consolidated structures, the resulting excess personnel capacity in peacetime combined with the ability to permanently equip FSLs with their full wartime tester requirements creates peacetime staffing location flexibility. The flexibility is produced by different ways in which the Air Force could assign "excess" peacetime personnel. This flexibility would allow the Air Force to balance deployment requirements with other personnel issues.

In recent years, small-scale contingencies and boiling peacetime operations have generated continuous deployments, resulting in what the Air Force has termed personnel turbulence. If consolidated structures can eliminate deployment requirements for these operations, the effects of deployment on F-15 AIS personnel would be minimal. By permanently assigning about one-fourth of "excess" peacetime personnel to FSLs in USAFE and PACAF West (regardless of consolidated structure), these FSLs could handle up to two AEFs in addition to the forces they normally support. This would eliminate all personnel deployments except for MTWs.

Determining assignments for the remaining three-fourths of the excess personnel would involve tradeoffs between the size of the CONUS rotation base and MTW deployment requirements. Placing all of these personnel forward in FSLs would minimize MTW deployment requirements. Figure 5.7 illustrates how this plan would function. The top two rows of numbers in this figure show the TISS tester requirements by region in peacetime and wartime. In "boiling peace" there is a requirement for 13 worldwide, but for two MTWs the Air Force needs 19. The maximum of these two numbers, 19, sets the minimum inventory of TISS testers that the Air Force needs. This is the minimum number of testers required at the maximum projected worldwide operating tempo. We note, however, that during peacetime the requirement is for seven in CONUS and during wartime it is for eight each in PACAF and USAFE. The sum of these individual-location peacetime/wartime maximums is 23. Since 23 is less than the current Air Force inventory of 25, there is no marginal equipment cost for permanently equipping each of the three regions with their maximum projected requirements for all anticipated scenarios, because the existing testers are a sunk cost. By doing this, the Air Force would eliminate the need to move testers when shifting from peacetime to wartime. Over the long term, however, personnel costs are variable, so it is possible to produce cost savings by reducing the personnel requirement below the existing personnel "inventory." Therefore, the Air Force should have enough avionics personnel to meet the maximum projected simultaneous demand of 19 sets of crews, but no more. A policy that would minimize MTW deployment requirements would permanently assign the "excess" peacetime personnel to the FSLs. Since CONUS has a maximum demand of seven sets of crews, this leads to assignments of six sets of crews at each FSL, three of which are "excess" during peacetime at each FSL. Then, in the event of an MTW in either AOR, two CONUS crews would augment the appropriate FSL. Because CONUS aircraft

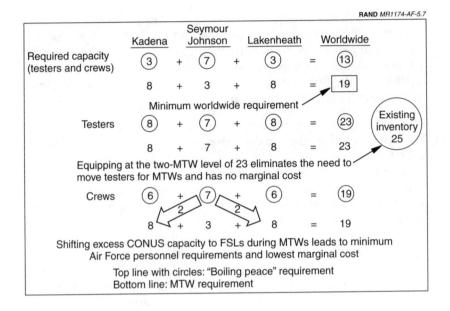

RAND *MR1174-AF-5.7*

**Figure 5.7—Wartime Augmentation of FSLs**

deploy during MTWs, CONUS capacity temporarily becomes excessive. Figure 5.7 indicates that if two MTWs occurred simultaneously, there would no reason to keep more than three sets of crews in CONUS.

Such a peacetime assignment policy would eliminate AEF-only deployments and slash MTW deployment requirements. Figure 5.8 shows the resulting maximum personnel deployments needed to augment FSLs during two simultaneous MTWs. Yet while this policy minimizes deployment requirements, it would create an unstable rotation base, as less than one-third of the force would be permanently stationed in CONUS.

Placing the excess capacity (again above AEF requirements) at the CSL instead would position two-thirds of the force—13 sets of crews in the TISS two-FSL example—in CONUS, producing an acceptable rotation base. This would result in three sets of crews for TISS at regional FSLs, as indicated by the required "boiling peacetime" capacity of three at Kadena and Lakenheath in Figure 5.7. In comparison

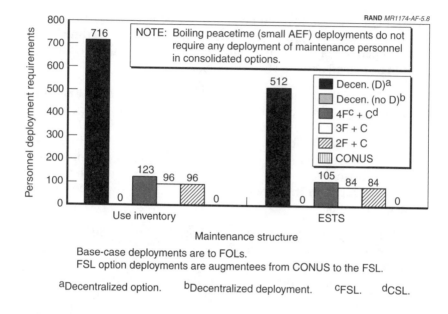

Base-case deployments are to FOLs.
FSL option deployments are augmentees from CONUS to the FSL.

[a]Decentralized option.    [b]Decentralized deployment.    [c]FSL.    [d]CSL.

**Figure 5.8—Minimum Two-MTW Deployment/FSL Augmentation Requirements**

to placing this capacity at FSLs, this would double or triple MTW FSL augmentation requirements and could possibly require some augmentation for medium-size contingencies. If MTW and medium-size operations remain infrequent, however, this option deserves serious consideration, because it would still eliminate the need for deployments for small-scale contingencies, drastically reducing the frequency of deployments requiring the deployment of avionics intermediate maintenance personnel. In fact, under this approach the only deployments that would have required avionics intermediate personnel to deploy during the 1990s would have been Desert Shield/Storm and ONA.

The potential use of the National Guard or the Air Force Reserve generates other options. F-15 avionics maintenance personnel could be kept at a level sufficient to accommodate the demands of two AEFs during peacetime, yielding 13 sets of crews in the TISS example. The remaining personnel necessary to support two MTWs could be reserve personnel who would be called as needed. This would eliminate active-duty personnel deployment requirements. A drawback is that all personnel turbulence induced by unpredictable deployment would be placed on reserve personnel.

Regardless of the manning options, however, we see that all of the consolidated options eliminate the deployment of avionics intermediate maintenance personnel for most contingencies, and of course the decentralized-no-deployment option completely eliminates personnel deployments. The question then becomes how important this benefit would be. To answer that, we need to examine the effect this might have on the retention of avionics intermediate maintenance personnel.

RAND researchers James Hosek and Mark Totten recently completed a study for the DoD in which they estimated the effects of long (greater than 30 days away from home) and hostile-location (regardless of length) deployments on the reenlistment rates of first-term and early-career personnel.[9] Their theory posits that each person who enters the service has his or her own preferred level of deployment frequency, intensity, duration, and risk profile. Their preferred profile is the one that they believe will maximize their satisfaction, or utility, with their job in the military. When they choose a service and occupation, among other factors, they will thus attempt to choose a deployment profile as close as possible to their preferred profile in order to maximize their total utility.

During their tour of duty, these personnel will experience an actual level of deployment that, if different from their expectations, will modify their beliefs about the deployments they should expect if they remain in the service. This actualization will also show them what deployments are really like, which may change their utility maximizing deployment profile. These changes will in turn affect their expected satisfaction from reenlisting and may thus influence the reenlistment decision.

---

[9]The research is described in detail in Hosek and Totten (1998). "Early career" is defined as the period after the first term and up to the tenth year of service. For the most part, early career covers the second reenlistment decision. For some personnel, however, early career will include a third reenlistment decision.

Hence, what we need to try to judge are the deployment preferences of personnel in the occupation of avionics intermediate maintenance. The personnel policy and structure option that most closely produces deployment requirements reflecting this preference will have the best effect on retention. We can make some inferences about the preferences of avionics personnel by examining the deployments that have been experienced by avionics personnel and the effects these deployments have had on retention.

Hosek and Totten added variables reflecting the utility of long and hostile deployments to a model estimating the probability of reenlistment, and they then examined the implications of long and hostile deployments on reenlistment for first-term and early-career personnel for each of the services. In general, they found that some deployment actually increases reenlistment rates. This seems consistent with the belief that those who enter the service seek a certain degree of adventure that deployment provides. However, increasing the length and frequency of deployment was found to reduce this effect—suggesting that, in essence, there can be "too much of a good thing." Experiencing one or more episodes of hostile deployment also reduced the positive effect on reenlistment probability.

The findings for the Air Force alone, however, were somewhat different. In this case, having at least one long or hostile deployment slightly decreased the enlistment probability of Air Force first-termers by about 1 percent with deployment being a positive factor for only 44 percent of personnel.[10] This suggests that recent deployment levels have, on average, come close to meeting preferences for those who enlisted. However, early-career personnel with at least one long or hostile deployment, were 10 percent more likely to reenlist than those with none, and deployment was shown to be a positive factor for 99 percent of this group—results that are both similar to those of the DoD as a whole. This suggests that those who reenlisted after the first term tended to be those who desired some degree of deployment. Unlike the general results, however, adding an episode of long or hostile deployment to those in either Air Force group who had already had at least one deployment did not change

---

[10]For the Army and Marine Corps this figure is 90 percent, suggesting that they attract many more people who desire deployments.

reenlistment probability. For those with at least one episode of long or hostile deployment, reenlistment probability was found to be lower if at least one episode was hostile—although for early-career personnel, it was still higher than if they had had no episodes of long or hostile deployment. So while early careerists tend to desire at least some deployment, as a group they would prefer that it be of a nonhostile nature.

What does this mean for the policy options considered in this analysis? It suggests that the options that reduce deployment requirements will not do much to affect the retention rates of first-term enlistees and may even reduce the reenlistment rates of early careerists. This is difficult to judge, however, because reducing deployments in the first term could result in a different set of personnel who choose to reenlist, and this, in turn, would produce a potentially different reaction to deployments on the part of early careerists. However, hostile deployments are consistently less frequently preferred than long, nonhostile deployments. So to the extent that personnel continue to deploy but do so to FSLs in safe locations rather than to FOLs in hostile areas, deployment should have a neutral to positive effect on retention.

From a personnel retention standpoint, this analysis would seemingly favor an option in which avionics personnel do deploy, but in which deployment is to nonhostile areas. This fits well with the FSL-inclusive options. It also suggests that if they are used, FSLs should be manned only at the minimum required peacetime level for each region—a level insufficient to accommodate small-scale AEFs. In the event of AEFs, a contingent from the CSL could then augment the appropriate FSL. With all those potentially deploying at the same CONUS location, this would further the EAF aim of leveling the deployment burden as evenly as possible among units and personnel.

The results for Navy and Marine personnel provide some insights into the potential effects of the EAF's scheduled deployments on retention. Both of these services carefully plan their long deployments, and both—especially the Navy—have been able to limit hostile deployments primarily to their scheduled periods of deployment. The retention analyses for these services suggest that this careful scheduling allows people to form more accurate expectations about deployments, which increases the likelihood that their expectations will align with actual outcomes.  Put another way, under these conditions expectations are more likely to be fulfilled and there will be fewer adverse outcomes because the service member has experienced either too much or too little deployment.  In effect, people know what they are signing up for, at least in terms of deployments—which helps the Navy and the Marines attract people who want the kind of experience they offer.

Hosek and Totten recognize that the increased workload that those remaining at home during deployments often experience may also affect retention.  To date, however, metrics to measure the effect of this workload have not been developed, so the extent to which reducing deployments improves the lives of those remaining at home and the effect this has on retention are unknown.

As mentioned at the beginning of the chapter, an alternative to improved personnel retention would be to find alternative sources of personnel.  The adoption of any structure that eliminates deployments to FOLs would increase the Air Force's flexibility in this regard. Should the Air Force decide to seek other sources of maintenance personnel, we note that eliminating deployments or keeping them limited to FSLs would enable the use of contractors, government-employed civilians, or allied partnerships.

## MINIMIZE DEPLOYMENT FOOTPRINT

The Air Force is working to understand the forces and deployment times needed to win the "halt phases" in future MTWs.  One key element in halt-phase operational success lies in using early airlift to carry combat forces to theater.  This places a premium on reducing deployment footprint.

With the current AIS, consolidation reduces initial airlift requirements by 18 to 60 C-141 load-equivalents for an MTW (see Table 5.1). The range comes from packing options.  If the AIS is transported in portable shelters, then it would take six C-141s to support each F-15 squadron with the current testers.  If the testers were packed in crates and the shelters left behind, an AIS for one squadron would require only 1.8 C-141s.  However, units need portable shelters to provide a clean working environment if proper facilities are unavailable, as is often the case at FOLs.

**Table 5.1**
**AIS Deployment Footprint**
**(Reduction in C-141 equivalents required for AIS**
**deployment current policy versus no deployment)**

|  | One MTW | One-Squadron AEF |
|---|---|---|
| Current AIS | 18–60 | 2–6 |
| ESTS configuration | 4–12 | 1–2 |

NOTE: Ranges reflect whether or not AIS deploys with shelters. Left number: deployment without shelters (testers only). Right number: deployment with shelters.

While consolidation with the ESTS configuration would also reduce up-front airlift requirements, as shown in Table 5.1, the reduction is much less because the ESTS footprint for each tester is only one-fifth that of what it is for the current testers, and each ESTS replaces up to five testers. Much of the airlift requirement for the ESTS configuration is actually for the TISS and antenna test stations. This again illustrates that there are often alternate means of reaching objectives. Both policy and technology alternatives can reduce the deployment footprint.

In exchange for reducing up-front airlift, options relying on resupply need guaranteed daily transportation to move spare parts. However, the total MTW requirement averages less than one pallet and only about 1000 pounds per day. This sustainment requirement might even be offset by a reduction in the required resupply of other items to the FOL. When people deploy to an FOL, they must be supported with food, water, fuel to run their vehicles, and other basic necessities. If the FOL is in a developed area, this materiel might be available locally. However, some austere locations might require that this materiel be brought in on a continual basis.

## MINIMIZE OPERATIONAL RISKS

### Resupply Risk

If the actual resupply times (through either local maintenance or nonlocal maintenance and transportation) do not meet the performance assumptions used to set spare-parts levels, then aircraft availability is likely to drop below operational goals.

## Flightline Reliance on Local Maintenance Resupply in Decentralization

In a decentralized structure, the greatest performance risk is tester downtime, because local maintenance provides the bulk of resupply to the flightline. The fewer the number of collocated strings in a location, the greater the probability that none will be available and the greater the variability in downtime. This leads to increased maintenance queues and hence to lower aircraft mission capability rates.

Figure 5.9 illustrates the risk associated with maintenance in a decentralized structure, especially when one tester string deploys in support of one squadron. Each of the two series represents six months of tester availability for one LRU for two different tester types (i.e., LRU A on the METS and LRU B on the computer test station) at Lakenheath. Lakenheath has three of one of the tester types and only one of the other. The reliability for both LRU-tester combinations is approximately 85 percent. The light triangles in Figure 5.9 are an example of the one-tester case and the dark diamonds represent the three-tester case. The y-axis represents the number of hours available each day per tester. The smoother this number, the smoother the daily supply of tester time and the more certainty a shop has in planning and executing maintenance.

In the one-tester case, the tester is either up or down, signifying that either 24 hours or no hours are available. Moreover, the downtime in this sample comes in stretches as long as ten straight days and, in the event of a difficult tester repair with hard-to-get parts, can exceed a month. If this happened during an AEF deployment, no repair could occur for the affected LRUs, and thus no resupply would take place without special intervention such as emergency setup of a distribution channel to the FOL. We term this single-string risk.

In the three-tester case, there were no days with zero hours available and only two days with two testers down. Nearly all days had two or three testers up. In effect, having three testers with a low level of reliability is like having two testers with very high reliability. The AIS can count on having at least two testers available almost every day.

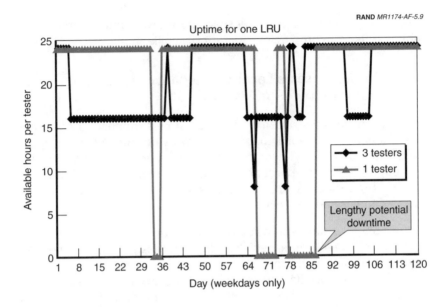

Figure 5. 9—Maintenance Single-String Risk

## In Consolidated Systems and the Decentralized-No-Deployment Structure, Resupply Depends on Establishment of a Theater Distribution System

Whereas the risk resulting from maintenance downtime in a decentralized structure is generally associated with resupply problems for a small set of LRUs (the set of LRUs tested or testers that experience lengthy downtime periods), problems associated with resupply from other locations are likely to affect all LRUs. Such problems would most likely manifest themselves as small across-the-board delays rather than no supply at all for a small set of LRUs.

The total daily MTW resupply requirement for a consolidated system is relatively low, representing only a small fraction of a C-130 load, but it is essential for the system to meet operational requirements. The graph in Figure 5.10 shows an example of what could happen if resupply planning time were not achieved. Consider a scenario in which the Air Force plans an RSP for a 12-PAA package with a DSO of

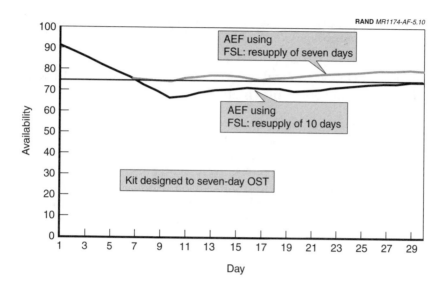

**Figure 5.10—Distribution System Resupply Risk**

75 percent on day seven based on a resupply time of seven days. The top light line in Figure 5.10 is the expected mission-capability rate of a 12-PAA AEF deployment over 30 days. From a low of 75 percent at day seven, the end of the period when the AEF is relying solely on the RSP, availability increases to 80 percent on day 30.

At the same time, the bottom, darker line in Figure 5.10 shows what could happen if a delay in establishing a theater distribution system in a new area of operations—in this case just three days—were to occur. Through day seven the picture looks the same, because in both cases the aircraft are relying on the RSP only. However, if system planning is based on resupply beginning on day seven and it does not arrive until day ten, then availability will continue to decline to a low of 65 percent on day ten as the RSP "unexpectedly" begins to run out of some needed LRUs after seven days. The gap between planned daily availability and actual availability will gradually close, but even by day 30, the availability will be up to only 75 percent—still 5 percent below what it would be with a resupply time of seven days.

If actual resupply time were to lengthen even more, aircraft availability would further deteriorate.

## Potential Delays in Establishing "Resupply" Also Pose Risk in the Decentralized-Deployment Structure

The decentralized-deployment structure assumes that testers will deploy and be operational by the third day of F-15 flying operations from the FOL. Any delay in establishing "resupply" through local repair is akin to a delay in establishing a theater distribution system for the consolidated options and thus yields results similar to those in Figure 5.10. For example, if it took six days to bring all of the testers on line, availability would drop about 10 percent below the expected level.

### Inventory Risk

For a variety of reasons (all beyond the scope of this report but currently being worked intensively by AFMC and the Air Staff), there have been times when the AFWCF has not fully recovered its costs and therefore the repairs and procurements required to maintain serviceable inventory levels have not been funded. This can leave RSPs short of required serviceable stocks, since RSP authorizations represent what should be on hand. For POS, the impact is more difficult to determine, since the authorized POS position includes serviceable inventory, in-transit assets (both serviceable and unserviceable), and assets due in from maintenance, whether on base or at a depot. If the sum of these assets is less than the authorization, then there could be an unfunded shortfall. This would be the case if unserviceable assets were not inducted into maintenance for purely financial reasons (as opposed to repair capacity or other reasons) or if required procurements were not placed.

Even in today's decentralized structure, any shortfalls in required RSP levels on deployment pose a risk to the Air Force. As the gap between requirements and on-hand inventory increases, the number of demands that cannot be immediately satisfied during a deployment is likely to increase, resulting in lower aircraft availability. Without local repair in the consolidated structures, aircraft availability relies solely on local inventory and resupply, thereby magnifying the risk posed by any RSP shortfalls.

POS shortfalls also pose a risk, but in a different form. POS gaps have an effect to the extent that aircraft are not mission capable at the time of a deployment order because they have been waiting for parts that should have been repaired or procured but were not. This, however, would create a problem only when the number of available aircraft is less than the deployment requirement.

## ENHANCE FLEXIBILITY

Over the last several years, Lakenheath has provided FSL-like support both to AEFs in SWA and for its own aircraft deployments.

The period of October 1998 to June 1999, which covers ONA, provides an interesting view of FSLs. During this period there were seven movements of Lakenheath F-15 squadrons or portions thereof, as depicted in Figure 5.11. Typically (doctrinally), these would have required an avionics test-string move to accompany each deployment. Since all of the units were supported from the Lakenheath FSL, however, no support equipment moved—saving 440 pallets or about 34 C-141s.[11] This "freedom of movement" created a "flexible" wing that made it easier for the Air Force to move F-15s around Europe and SWA as the dynamic political situation changed. It thus seems that use of consolidated regional repair increases the Air Force's ability to continuously adjust the position of forces to meet the needs of a dynamic environment.

In some situations, however, consolidated regional repair that does not deploy may actually limit flexibility or at least require backup deployment plans. Essentially, an FSL must have access to commercial transportation to its supported FOLs or be within C-130 range (refueled range or accessible through an en-route stop) of supported FOLs. Thus, while FSLs in USAFE and PACAF would allow the Air Force to effectively cover a large portion of the world, parts of Southern Africa and Central Asia might be difficult to support. If

---

[11]The FSL was called a Centralized Intermediate Repair Facility (CIRF) by Lakenheath and USAFE.

RAND *MR1174-AF-5.11*

Lakenheath CIRF enabled greater operational flexibility and reduced deployment burden

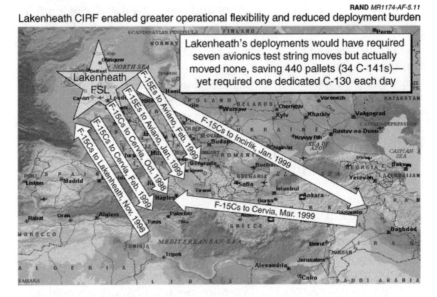

Lakenheath's deployments would have required seven avionics test string moves but actually moved none, saving 440 pallets (34 C-141s)— yet required one dedicated C-130 each day

**Figure 5.11—Lakenheath F-15 Unit Movement from October 1998 to March 1999**

most of the key areas of concern for potential F-15 employment are within the support range of FSLs, this is a minor issue that can be handled by setting up contingency plans to deploy part or all the FSL in the event of the need for F-15s in other areas.

## SPIN-UP TIME

By design, each of the structures removes the AIS from the critical path for commencing operations in a deployment. The decentralized-deployment structure is designed to provide sufficient inventory of RRR LRUs to sustain aircraft until the AIS is operational, as well as enough RR LRUs and SRUs to meet demand over the first 30 days. The consolidated structures and the decentralized-no-deployment structure assume that the distribution system between support locations and FOLs will be operational from the start of a conflict. Therefore, there is no spin-up time difference between the options. Potential failure to meet these assumptions is addressed in the risk discussion above.

# Chapter Six

# ADDITIONAL OPPORTUNITIES TO IMPROVE LOGISTICS SYSTEMS WITH CONSOLIDATED STRUCTURES

To this point, our analysis has assumed that processes and process assumptions would remain constant for all repair options. With the consolidated options, however, processes can be redesigned to either enhance capability or reduce the estimated resource requirements while maintaining performance.

## INCREASED TESTER UPTIME AND AIS THROUGHPUT

In computing the number of testers required for each support option, we held tester uptime constant. It is likely, however, that consolidated shops can achieve higher average uptimes and better throughput.

First, having several test stations of the same type would allow the AIS to dedicate some testers to high-demand LRUs, which would reduce setups. Reducing setups decreases total repair time by the amount of time it takes to change from one setup to another. Setups are also a major cause of test station failures, so reducing setups should also increase tester uptime.

Concentrating testers in fewer locations would ease the burden of providing tester spare parts. Consolidating the spare-parts inventory now distributed across ten home-base operating locations would increase tester spare-parts stocks at centralized locations without any investment, thus improving potential inventory effectiveness. If a shop still found itself short of tester spare parts, having more collocated testers of the same type would increase the opportunity for tester cannibalization. The shop would be able to leverage a "can" tester to keep multiple testers going rather than just one or two. Thus, regardless of the number of collocated testers, a shop should be able to limit the number of testers down for the same LRU to one or two, whether the shop has three testers or six. Finally, consolidating tester assets would also consolidate tester maintenance personnel. Each location would then have more experienced and skilled technicians available for troubleshooting and for providing on-the-job training of other personnel.

The excess testers in the consolidated options also offer spare-parts protection through spare testers. Even after setting up the maximum number of testers in each FSL and CSL, a small number of excess testers would remain. In the TISS crew and equipment-positioning example discussed earlier, for example, worldwide inventory is 25—two greater than the total system need of 23. FSLs could use these as spare testers in wartime (they are unlikely to be necessary in peacetime because tester capacity is at about 150 percent of peacetime demand).

Besides increasing effective capacity through better tester uptime and reduced setup time, consolidation may decrease actual repair time. Keeping testers and people together in one location can improve troubleshooting of LRUs by allowing for the comparison of results across stations when diagnostics results are not clear, as well as by pooling technical expertise. Improving troubleshooting should lead to a higher first-time repair success rate. Similarly, centrally located LRUs that are unserviceable and awaiting maintenance or parts provide greater cross-cannibalization opportunity, which could reduce average AWP time in the AIS.

## REDUCING MAINTENANCE WAIT TIME

When personnel and tester levels for a shop and given test station type are set, a tradeoff is made between capacity utilization and total in-shop time that includes both queue time and repair time. Currently, Air Force spare-parts calculations are based on an average base repair time of four days. Because the actual repair time for an LRU averages approximately one shift, much of the in-shop time must be spent waiting for repair. In effect, the maintenance system pursues high-capacity utilization (over 90 percent) in order to reduce maintenance resource costs. But this has a cost in terms of increasing the length of the spare-parts pipeline.

Under the current decentralized system, such capacity planning is rational because it would take a significant increase in capacity per location (with the resulting drop in capacity utilization) to dramatically reduce queue times. As the number of collocated testers of the same type increases, however, the effect of high-capacity utilization on queue time decreases. In the tester quantity estimates presented earlier, we held the queue time constant for a total base repair cycle time of four days. In the consolidated options, it is possible to increase capacity just slightly (decreasing capacity utilization) and achieve a significant reduction in total repair time.

Figure 6.1 illustrates the consolidation effect of "load" smoothing and helps explain why consolidation makes a tradeoff between capacity and queue time advantageous. When servers in a queuing system are collocated—as when testers are consolidated at an FSL—demand is smoothed across the servers. Thus, collocated servers can handle demand more effectively than the same number of servers operating independently. The demand represented in this figure is that of five squadrons for LRUs on a given tester type. The right-hand bars of each pair represent decentralized maintenance in which independent shops support each squadron. The left-hand bars represent consolidation of the five testers in one FSL to support all five squadrons. Starting from an empty system at day one, demand variability results on day ten in an expected maintenance backlog in the decentralized system that is more than double the expected backlog of the consolidated system (with tester availability and repair time treated as constants). While both systems have the same total capacity at a level nominally sufficient to handle total demand, demand variability produces backlogs. Adding in variability due to repair time and tester uptime would further increase backlog differences between the two systems.

In computing tester and personnel requirements, we set the backlogs (in effect, the queue time) to be equal for all repair structure options. This means that the two sets of bars on this chart would on average

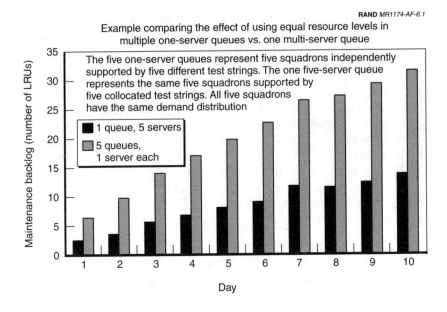

RAND *MR1174-AF-6.1*

Example comparing the effect of using equal resource levels in
multiple one-server queues vs. one multi-server queue

The five one-server queues represent five squadrons independently
supported by five different test strings. The one five-server queue
represents the same five squadrons supported by
five collocated test strings. All five squadrons
have the same demand distribution

■ 1 queue, 5 servers

▨ 5 queues,
   1 server each

Maintenance backlog (number of LRUs)

Day

**Figure 6.1—The Consolidation Effect of Load Smoothing**

be equal in height. To achieve equality, one must either increase the
number of testers at each decentralized location or decrease the
number of testers in the consolidated locations.

What would it take to reduce the size of the spare-parts pipeline by
decreasing maintenance queue time? In Figure 6.1, adding one test
station to the consolidated option would result in zero backlog at day
ten. Achieving the same result in the decentralized case would require a doubling of the number of testers to ten. Adding capacity to
reduce the wait time in a decentralized system with variable demand
can be prohibitively expensive.

Figure 6.2 shows the results of increasing capacity to reduce queue
time for different support structures. The left bars of each set represent the additional testers and personnel needed to reduce the base
repair cycle time to about one day. The middle bar is the resulting
decrease in inventory costs, and the right bar is the total change in
cost. As consolidation increases, the costs to reach the desired capacity decline and inventory savings climb rapidly.

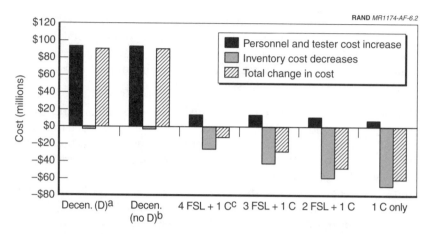

Figure 6.2—Reducing Spare Parts by Increasing Maintenance Capacity

The inventory decrease for the decentralized case is extremely small because we treat current inventory as a sunk cost, whereas the savings for the consolidated cases come from reductions in the additive buy and repair requirements. As in the decentralized case, the shorter existing pipelines, the smaller the potential savings from reducing pipeline length, because operating and safety stocks are smaller.

## REDUCING OSTs

There may also be opportunity in the consolidated structures to reduce the pipeline by improving the retrograde and shipment times between FOLs and FSLs. These pipeline times are probably easier to reduce than those between bases and depots, because a new system would be less constrained by existing systems, procedures, and structures. The Air Force could design a new retrograde and order-delivery system from the ground up. Communication could be point to point from operating locations to repair activities through the Internet, and delivery could be made right into shops collocated with small avionics warehouses. Together with simplified warehouse management, these possibilities should enable reductions in order processing and receipt take-up times compared to the depot system. This is akin to the opportunities manufacturers often gain when they build greenfield sites rather than overhauling existing facilities: They find it easier to implement new ideas from the start of an operation than to modify existing processes.

Further, it would become the responsibility of the FSLs and FOLs to manage retrograde and delivery times. This would remove the barriers of "functional silos" from resupply and create single-point ownership of the maintenance process—a process that would include resupply to FOLs. Consequently, it may be more appropriate to compare the decentralized systems using baseline resupply times with consolidated systems using fast rather than baseline times. Unless the depot system and the multiechelon information processing requirements were to be revamped, the consolidated systems should be able to achieve faster resupply times than those for comparable decentralized resupply segments.

Lakenheath's transportation success during ONA provides support for the potential ability of FSLs to set up and manage a highly effective distribution system. Basing its action on prior experience with FSL support, Lakenheath went to great lengths to plan distribution to FOLs for a potential conflict such as ONA. In addition, the supply and transportation staffs directly managed distribution planning and execution. As a result, they were able to achieve OSTs for Mission Impaired Capability, Awaiting Parts (MICAP) requisitions from the FSL to FOLs at Aviano and Cervia, Italy, under the three-day faster time used in this analysis. The mean was just 2.3 days and the median 2.1, and 95 percent of shipments arrived within 2.8 days. Total OSTs for non-MICAPs were not available; however, transportation and receipt takeup times for all priorities from Lakenheath to the FOLs averaged less than 2.5 days. Thus, the OSTs for these shipments were likely to have been at least in the fast range of five days (giving 0.5 day each for order processing and materiel release and 1.5 days for pick-pack and move from base supply to base transportation).

## SHARING RESOURCES IN THE SYSTEM OF THE FUTURE

When a broader view is taken of the future, the transportation portions of the FOL/FSL and home-base/CSL segments may offer scale advantages. If consolidated repair of avionics occurs as part of a broader system of consolidated maintenance—e.g., including engines—then the various maintenance systems could share transportation resources. From an avionics standpoint, the low weight and cube requirements could drive the actual marginal cost virtually to zero if consolidation were to be widely adopted. For example, consolidated engine or LANTIRN (Low Altitude Navigating and Targeting Infrared Night) repair could require military airlift that a consolidated avionics maintenance system could share.

## A POTENTIAL STRATEGY FOR REDUCING MARGINAL INVENTORY INVESTMENT IN CONSOLIDATED STRUCTURES

The biggest cost driver in the consolidated options is the increased pipeline length from the flight line to the first echelon of repair. In decentralized structures, this pipeline is the base repair time, whereas for consolidation it includes shipment to and from the FSL or CSL. However, many of the parts in the consolidated structures that require an increase in total inventory are in sufficiently long supply to allow for such an increase at little (the cost to repair unserviceables) or no cost. A few expensive parts not in long supply contribute the bulk of the net cost. A part-by-part analysis reveals the high concentration of net procurement and repair costs among just a small number of LRUs. In fact, just ten LRUs account for about 90 percent of the net cost in the four-FSL/one-CSL option, and the other consolidated options have similarly high proportions of cost from just a few LRUs. The four-FSL/one-CSL list is provided in Figure 6.3, which provides the top 20 avionics LRUs ranked by net investment cost. The first column is the part national stock number (NSN), the second the net investment cost of the NSN, the third the cumulative net investment cost, and the fourth the cumulative net investment cost for those LRUs that can be tested on the ESTS; the fifth column provides the testers that can test each NSN.

Finding a way to reduce the increased pipeline requirement for a few of the NSNs that drive the net cost while minimizing increases in other costs could cut the costs of the consolidated options while preserving their benefits. Repairing some of these NSNs on base during peacetime would cut the increased inventory position requirement while circumventing the need for resource deployment. This would be cost-effective if the present value of the investment in additional testers and the personnel costs required for peacetime home-base repair were less than the reduction in spare-parts investment. In the consolidated options, however, there is excess tester inventory (and excess planned inventory of ESTS) that can be leveraged. First we need to examine which testers are needed to repair the high-net-cost LRUs. Figure 6.3 reveals that nine of the top 20 LRUs, which account for more than 100 percent of the net cost (some LRUs actually have a negative net cost in the consolidated options owing to safety stock reductions), can be repaired using the ESTS. This includes three of the top four, which together contribute more than 50 percent of the cost. Thus, the extra ESTS testers could be used at bases without

RAND *MR1174-AF-6.3*

| NSN | Net Cost | Cumulative Cost | Cumulative Cost (ESTS) | Testers |
|---|---|---|---|---|
| 5841013150646FX | $ 22,611,775 | $ 22,611,775 | $ 22,611,775 ESTS | Microwave |
| 6605012400136FX | $ 18,008,496 | $ 40,620,272 | $ 40,620,272 ESTS | METS |
| 6610003494184 | $ 10,284,396 | $ 50,904,668 | | Computers |
| 1270013841108FX | $ 9,582,536 | $ 60,497,204 | $ 50,202,808 ESTS | Microwave/displays/computers |
| 5996013451134EW | $ 5,676,911 | $ 66,164,115 | | TISS |
| 5996012428347EW | $ 5,635,346 | $ 71,799,461 | | TISS |
| 5841012348535FX | $ 3,982,648 | $ 75,782,109 | $ 54,185,456 ESTS | Microwave |
| 5895014124396EW | $ 3,453,022 | $ 79,235,131 | | |
| 5895014139798EW | $ 2,981,904 | $ 82,217,035 | | TISS |
| 5841013760002FX | $ 2,111,637 | $ 84,328,672 | $ 56,297,094 ESTS | Displays |
| 5985012778913FX | $ 2,066,148 | $ 86,394,820 | | EARTS ANT |
| 1270012368438FX | $ 2,061,052 | $ 88,455,872 | $ 58,358,146 ESTS | Displays |
| 5865012428918EW | $ 2,012,312 | $ 90,468,184 | | TISS |
| 5895012278102FX | $ 1,947,747 | $ 92,415,931 | $ 60,305,892 ESTS | Microwave/computers |
| 5841011007363FX | $ 1,700,136 | $ 94,116,067 | | EARTS/ANT |
| 6610010848224 | $ 1,632,696 | $ 95,748,762 | $ 61,938,588 ESTS | Displays |
| 6130012905835EW | $ 1,242,593 | $ 96,991,355 | | TISS |
| 5865012876182EW | $ 1,210,548 | $ 98,201,903 | | TISS |
| 5996010456276EW | $ 1,055,392 | $ 99,257,295 | | TISS |
| 5895011736012EW | $ 707,226 | $ 99,964,521 | $ 62,645,813 ESTS | Microwave/displays/computers |
| Total net buy + repair = $93,918,172 (the cumulative cost begins to decrease as one goes further down the list, because the net cost for some NSNs is negative). | | | | |

**Figure 6.3—Net LRU Spare Costs (Four-FSL + One-CSL) by LRU
(NSN is the national stock number for a unique part)**

collocated FSLs or CSLs. To take maximum advantage of this, we can go down the list of NSNs in Figure 6.3 and add LRUs to a local-testing list until we reach the daily capacity of one tester. This represents the most cost-effective one-ESTS capacity package for peacetime home-base repair. Then, as long as the present value of the associated personnel cost is less than the spare-parts investment cost "savings," this plan would reduce the cost of consolidated options.

Since there would be five excess ESTS in the four-FSL and one-CSL plan, an ESTS could be provided at the five bases with the most aircraft not collocated with a repair facility. If the CSL were to be at a high-volume CONUS base and FSLs were collocated with the F-15 bases in PACAF (Kadena) and USAFE (Lakenheath), then all but Edwards and Nellis AFBs could be covered (there are ten F-15 locations), thus cutting net investment by about $50 million. Since the testers would be excess to requirements, the only incremental cost would be people, which would come to a net present value of $8 million over eight years. This would effectively cut the cost of the four-FSL/one-CSL plan by about $40 million over eight years.

# Chapter Seven

# CONCLUSION

This chapter summarizes the key areas in which the various alternative structures affect the ability to meet EAF goals.

## DECENTRALIZED WITH AIS DEPLOYMENT STRUCTURE

The major advantages of the current decentralized structure in which the AIS deploys to FOLs are low relative cost and elimination of the need to immediately establish an effective theater distribution system upon initiation of a contingency operation in a new area. Yet another benefit is the "location flexibility" the structure provides; its performance should be roughly the same regardless of the remoteness from current Air Force locations.

The risks and disadvantages of this option, however, prompted the examination of alternative structures and have already caused many deploying units to informally modify their procedures. To meet operational objectives, the current structure requires not only a large number of trained personnel but possibly more than the Air Force currently has. As noted, our analysis model estimates the need for slightly greater requirements for TISS and ESTS testers and associated personnel than those included in the current Air Force plan. Further, these people may face continued, frequent deployments to hostile areas, which some believe has contributed to retention problems in the critical skill area of avionics technicians. While Hosek and Totten's research concludes that some deployment may actually have a positive effect on reenlistment, they also show that too many deployments may be "too much of a good thing." Their work further suggests that avoiding deployments to hostile areas that are some-times required by this option could improve retention.

In deployments, the AIS would consume valuable, constrained airlift capacity, which reduces the up-front lift available for combat forces. Once deployed, the AIS would often consist of only one set of testers, because AEFs frequently deploy in squadron-sized or smaller ele-ments. Resupply of LRUs then faces the single-string risk. Since spares planning assumes that the AIS will be operational on day three, having to ensure almost immediate operability of this complex equipment poses additional risk.

## DECENTRALIZED WITHOUT AIS DEPLOYMENT STRUCTURE

Modifying the current structure to eliminate deployment of the AIS would eliminate the deployment-induced airlift disadvantage of the current decentralized structure and facilitate dynamic changes in F-15 beddown locations. However, the total elimination of person-nel deployments could actually have a negative effect on personnel retention.

Besides reducing deployment footprint, eliminating the deployment of the AIS would remove the single-string risk because most home operating bases have multiple strings of testers (single-string risk is primarily a single-squadron or small AEF deployment problem). In-stead, the resupply risk would come from resupply through trans-portation from home bases that would depend on the establishment of a theater distribution system for contingencies in new areas of op-erations. Home-base repair during contingencies would also in-crease the importance of retrograde transportation, an area that has not received emphasis. Additionally, retrograde and shipping from many individual home bases to FOLs might stress the C2 system more than the traditional pipeline routes between FOLs (or home bases) and depots. During wartime, this would create a complex OCONUS-to-CONUS point-to-point system filled with low-volume channels. Related to the resupply risk is that posed by any gaps that develop in peacetime between RSP authorized and on-hand levels, because aircraft availability would rely solely on inventory and re-supply rather than on a combination of local repair and inventory.

Cost may be the biggest disadvantage of this structure, however. The need for an increase in the serviceable inventory of spare parts would require a one-time investment that makes this option more costly than the current structure.

## CONSOLIDATED STRUCTURES

As with the decentralized structure without deployment, consolidation would eliminate the deployment footprint concern associated with the current structure, thereby enhancing flexibility to meet dynamic F-15 beddown needs, and would also provide what is probably the most attractive personnel deployment profile: infrequent deployments to nonhostile areas. However, this option is less expensive than the decentralized-no-deployment option and approximately the same cost as the current structure with today's resupply times and testers or fast transportation and ESTS. And if some of the unique opportunities it offers—such as reducing repair time by reducing queue time through a slight capacity increase—could be successfully implemented, the total cost would drop. The option has the added advantage of reducing total personnel requirements, which would ease the strains on proper staffing of trained personnel.

Like the decentralized-no-deployment option, consolidated repair would also depend heavily on consistent retrograde and resupply availability through a theater distribution system, although command and control of these flows would be less burdensome. All retrograde from a region would flow to the same point, and resupply would emanate from that point. Likewise, with exclusive FOL reliance on local inventory and resupply, consolidated structures face the risk posed by any unfilled RSP holes that develop during peacetime.

Consolidating introduces another element of risk—implementation. Are the estimated resource requirements correct? What issues have not been anticipated? A detailed planning team and a test could help ameliorate this risk by improving the information available for a final decision. One implementation concern is how units and the FSLs/CSL would interact. It would be imperative to ensure that the consolidated repair locations treat the aircraft they support with the same sense of ownership and urgency fostered by decentralization. Similarly, units would have to be satisfied that the Air Force Materiel

Command (AFMC) would keep even higher RSP authorized levels filled.

## TEST OF AN F-15 REGIONAL REPAIR FSL

During Operation Noble Anvil (ONA), the air war against Serbia, the 48th Component Repair Squadron at RAF Lakenheath implemented the FSL repair concept as part of a system of FSLs set up by USAFE, thereby formalizing practices they had used on an ad hoc basis over the last several years. Lakenheath personnel successfully supported their own aircraft at FOLs as well as concurrent deployments to SWA using existing assets without deploying any AIS personnel or equipment. In fact, between October 1998 and March 1999, as tensions rose or eased, the wing supported by this squadron made seven different partial-unit deployments back and forth from Lakenheath to SWA and Italy without moving the AIS (depicted in Figure 7.1). Normally, Air Force policy would require that these deployments include the AIS, but since all of the units were supported from the Lakenheath FSL, no support equipment had to move. As a result, airlift requirements for these seven deployments were reduced by 35 C-141 sorties. More than any theoretical description of the flexibility that FSLs can provide in today's dynamically shifting environment, these operations demonstrated the advantage nondeploying maintenance structures confer in facilitating the repositioning of forces as quickly as political situations change.

The squadron also implemented plans for the Lakenheath avionics maintenance FSL to support an augmentation of F-15s from CONUS for ONA with just half the deployment footprint and personnel that would have been required had the deploying-wing AIS moved to the new FOL. In a permanent consolidated structure, even this limited equipment deployment would not have been required, because the equipment would already have been in place; thus, only a limited number of personnel would have had to deploy. In exchange for the reduction in deployment airlift, the FSL had to rely on a steady flow of transportation to provide resupply to the operating locations.

Lakenheath logisticians used their prior experience, including that gained in the October 1998 deployment to Cervia, Italy (shown in Figure 7.1), to conduct transportation planning for providing support

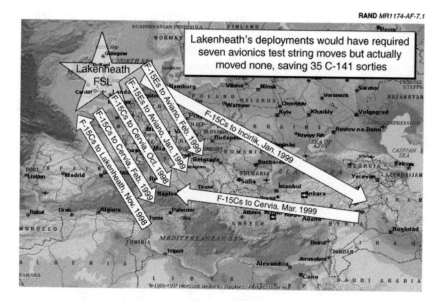

Lakenheath's deployments would have required
seven avionics test string moves but actually
moved none, saving 35 C-141 sorties

**Figure 7.1—At Lakenheath FSL Facilitation of Operational Flexibility
During a Time of Heavy Political Turbulence**

from an FSL. This enabled it to provide rapid and responsive
resupply of serviceable parts to FOLs from the start of ONA through
the intratheater distribution system and a Lakenheath-managed
"distribution system" that augmented the joint system. The Lakenheath distribution system was critical to the success of the operation.
Other Air Force FSLs established in support of ONA relied solely on
the joint intratheater distribution system but did not find it sufficiently responsive.

## CONCLUSION

Air Force leaders and F-15 avionics maintainers in USAFE, PACAF,
and ACC made it clear that they highly value the reduced deployment footprint, lower personnel turbulence, and flexibility that consolidation affords, and cost makes at least the four-FSL/one-CSL option feasible. However, the key disadvantage they cited with consolidation is the risk posed by the need to quickly establish a wartime
theater distribution system. As discussed earlier, Lakenheath suc-

cessfully established its own effective distribution system, but other ONA experience suggests that this result has not been consistently achieved. Also, given the difficulties the Air Force has sometimes had in keeping RSPs at authorized levels, some are reluctant to move to a structure that relies more heavily on local inventory than on local repair. Thus, before moving to any kind of structure that eliminates AIS deployment (consolidated or decentralized-no-deployment structures), AFMC, the Air Staff, and the major commands should agree on inventory replenishment policies and practices that produce a high level of confidence in the field.

In addition to continuing its ongoing review of financial policies and how they affect inventory replenishment, we recommend that the Air Force review current wartime theater distribution plans for consistency with potential EAF ACS operating procedures and then work as part of the joint community to modify them as necessary to address identified risks or performance gaps. Even if the Air Force then elects to retain the current structure, improving the wartime theater distribution system will reduce single-string risk by increasing the ability of the distribution system to provide backup support.

Assuming that the Air Force and joint community develop "reliable" plans for wartime theater distribution, however, we recommend the adoption of a consolidated network of regional repair locations to reduce deployment burdens and enhance flexibility if the Air Force continues to use the current testers. Such a network would provide more benefits than ESTS adoption at less cost. If the Air Force proceeds with ESTS implementation, consolidation would have a somewhat greater cost and would provide marginally fewer benefits than it would with the current testers. In this case, the Air Force would need to assess the value it places on the marginal reductions in personnel deployment requirements to hostile locations and in deployment footprint that a consolidated structure would provide beyond adopting ESTS alone. The decision should then come down to how these values compare to the associated spare-parts investment requirement.

# TESTER REQUIREMENTS MODEL

The tester requirements model calculates the minimum number of testers necessary at one location to support a given number of aircraft operating at a specified tempo subject to a maximum expected wait time. As shown in Figure A.1, the model consists of three components: (1) determining maintenance shop demand, which is the total tester time required per day for each tester type;

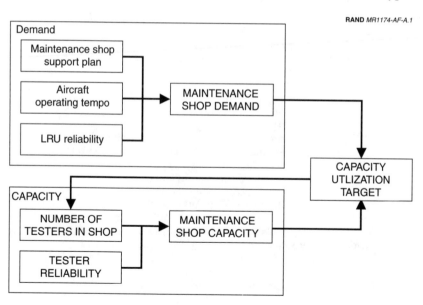

RAND MR1174-AF-A.1

Figure A.1—Tester Requirements Model Framework

(2) determining the daily capacity of each tester type for a given number of testers; and (3) adjusting the number of testers until the expected maintenance wait time is less than or equal to the target. The maximum expected wait time can be expressed as a set of maximum capacity utilization targets for different numbers of collocated testers. This appendix details the technical structure of the model.

## MODEL TERMS

The terms used to describe the model are outlined in Table A.1.

**Table A.1**

**Summary of Model Terms**

| Variable | Definition |
|---|---|
| $DD(x)$ | Daily demand on test station $x$ (hours) |
| $DD_i(x)$ | Daily demand on test station $x$ by LRU $i$ |
| $DDC_i(x)$ | Daily demand on test station $x$ by LRU $i$ for F-15Cs |
| $DDE_i(x)$ | Daily demand on test station $x$ by LRU $i$ for F-15Es |
| $FHC$ | Flying hours per day for F-15Cs |
| $FHE$ | Flying hours per day for F-15Es |
| $PTOIMDR_i$ | Peacetime supply system demand for LRU $i$ (LRUs removed/100 flying hours) (D041) |
| $NRTS_i$ | NRTS for LRU $i$ (%) (D041) |
| $\#REM_i$ | Number of maintenance removals at Seymour Johnson AFB and RAF Lakenheath over six months for LRU $i$ |
| $\#AWP_i$ | Number of repairs that were initially AWP at Seymour Johnson and Lakenheath over six months for LRU $i$ |
| $\#BCS_i$ | Number of BCS removals at Seymour Johnson and Lakenheath over six months for LRU $i$ |
| $\#REP_i$ | Number of successful repairs at Seymour Johnson and Lakenheath over six months for LRU $i$ |
| $AWP\%_i$ | Percentage of repairs initially AWP for LRU $i$ |
| $RDR_i$ | Repair demand rate (per 100 flying hours) for LRU $i$ |
| $BDR_i$ | BCS demand rate (per 100 flying hours) for LRU $i$ |
| $NDR_i$ | NRTS demand rate (per 100 flying hours) for LRU $i$ |
| $REPT_i$ | Repair test station time for LRU $i$ |
| $NT_i$ | NRTS test station time for LRU $i$ |
| $BCSTIME_i$ | BCS test station time for LRU $i$ |
| $n(x)$ | Number of testers of type $x$ |
| $Cfy$ | Maximum capacity utilization for $y$ collocated testers |
| $CU(x)$ | Capacity utilization of tester type $x$ |

## TEST STATION DEMAND

Test station demand is the number of test station hours required per day to test and repair all LRUs that use the given test station. Determining demand is a four-step process outlined below and in Figure A.2.

1. Determine the LRUs that use the given test station.
2. Determine the test station hours required per flying hour for both F-15Es and F-15Cs for each LRU.
3. Use the operational scenario's flying-hour requirements to determine the demand for each LRU.
4. Sum across all LRUs that use the given test station.

Each test-station type is designed to test a set of LRUs. The F-15 System Program Office (SPO) maintains a list with the capabilities of

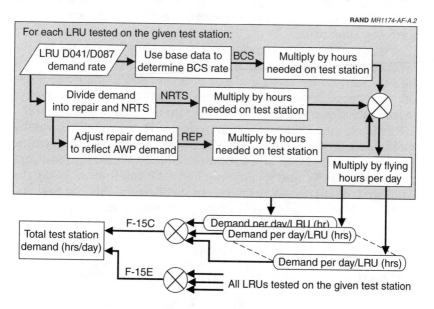

Figure A.2—Test Station Demand Calculations

each test station. While most LRUs are only testable on one test station, some are testable on multiple test station types (within the same test string configuration). For example, several LRUs are testable on the display, microwave, and computer test stations. With the model used in this analysis, users can readily change the station on which each LRU is tested. For purposes of this analysis, such LRUs were tested on the test station with the lowest total demand from other LRUs in order to balance demand.

After the list of LRUs for each test station has been determined, the demand created by each LRU must be ascertained. LRU removals from aircraft result in three types of demand: repaired on base, not repairable on base (NRTS), and false failures (BCS). The Air Force supply system captures removals that result in on-base repairs and NRTS; the information system that records these data is called D041. D041 provides two data elements: removals per 100 flying hours (PTOIMDR) and the percentage of these that are NRTS. Multiplying them then provides the NRTS event rate and one minus that rate yields the repair event rate. However, repairs may potentially visit the test station twice. For example, if the test indicates the need for a part that is not in stock and the shop does not have another LRU available for cannibalization, then the maintenance technician must take the LRU off the tester and tear the tester down (there are different setup configurations for different LRUs). When the part comes in, the technician will then have to set up the station again and the LRU will consume more test station time. While the LRU is waiting for the part, the status is termed AWP. Manual data collection efforts from Seymour Johnson and Lakenheath recorded LRU removals, repairs, NRTS, BCSs, and repairs that had AWP conditions. Using the AWP and repair totals provides an AWP event rate for repairs. Adding this to the repair event rate provides the total repair event demand rate on the tester.

$$AWP\%_i = \#AWP_i / \#REP_i$$

$$RDR_i = (1 - NRTS_i) * PTOIMDR_i * (1 + AWP\%_i) \quad \text{repair demand rate}$$

$$NDR_i = NRTS_i * PTOIMDR_i \quad \text{NRTS demand rate}$$

Similarly, the ratio of BCSs to the sum of base repairs and NRTS serves as the relative BCS event rate based on the PTOIMDR.

$$BDR_i = PTOIMDR_i * \#BCS_i / (\#REM_i - \#BCS_i) \qquad \text{BCS demand rate}$$

For some LRUs used in the model, the PTOIMDR was not available. In these cases, the model estimates PTOIMDR through a linear equation developed from a regression with PTOIMDR as the dependent variable and base removals (from Lakenheath's and Seymour Johnson's manual data collection systems) as the independent variable. The next section describes this estimation technique in more detail.

$$PTOIMDR_i \text{ (est)} = \#REM_i * 0.002254$$

The repair and NRTS times are estimated as twice the BCS time (BCS time includes test documentation time, test station setup time, test time, and test station teardown time).[1]

$$REPT_i = 2 * BCSTIME_i$$

$$NT_i = 2 * BCSTIME_i$$

Multiplying event durations by event rates provides the demand rates (hours of demand) per flying hour. This is done for the set of F-15C LRUs and that of F-15E LRUs to produce two kinds of demand rates:

$$DRC_i = (RDR_i * REPT_i + NDR_i * NT_i + BDR_i * BCSTIME_i) / 100$$
$$\text{for all LRUs(i) on F-15Cs}$$

$$DRE_i = (RDR_i * REPT_i + NDR_i * NT_i + CDR_i * BCSTIME_i) / 100$$
$$\text{for all LRUs(i) on F-15Es}$$

Multiplying by daily F-15C and F-15E flying hours and then adding the F-15C and F-15E daily demands provide the daily demand for each LRU on tester $x$. If a wartime scenario is being modeled, the flying hours should be multiplied by the appropriate deceleration

---

[1] For repairs, this time includes running through the test until the tester indicates a problem, making the repair, and then running through the complete test to verify that the LRU is serviceable. For NRTS events, significant time is often spent trying to make the repair before the LRU is declared NRTS. Personnel at several shops considered the estimates of twice the BCS time reasonable. Actual total time on test station was not available.

factors.[2]  Previous research has determined that avionics LRU failure rates are not linear with respect to sortie durations.[3]  The Air Force employs the deceleration factors to estimate how the failure rates per flying hour would change at the predicted wartime sortie duration.

$$DDC_i (x) = DRC_i * FHC * \text{deceleration factor (if a combat scenario)}$$

$$DDE_i (x) = DRE_i * FHE * \text{deceleration factor (if a combat scenario)}$$

$$DD_i (x) = DRC_i * DRE_i$$

Summing across all LRUs provides the total daily demand on tester $x$.

$$DD(x) = \Sigma \, DD_i (x)$$

## REGRESSION OF NUMBER OF REMOVALS AGAINST PTOIMDR

This section describes the results delineated in Table A.2 of a regression of the number of removals against PTOIMDR.  The resulting linear equation is used to estimate PTOIMDR when it is not available through D041 data.

The F-test and R-square calculations are typical measures of goodness of fit of a regression model.  The F-test significance implies that we can accept the hypothesis of a linear relationship between on-base removals and PTOIMDR with greater than 99.999 percent confidence.  The R-square of 0.21 indicates that the actual number of on-base removals at Lakenheath and Seymour Johnson for this set of LRUs explains 21 percent of the variation in PTOIMDR.  In terms of goodness of fit, this regression model was the best of several different models employing other variables available from base data.  While the R-square is relatively low, indicating significant differences between the actual and predicted values, the absolute removal rates of the set of LRUs requiring the use of this model are very low (all of the LRUs without PTOIMDRs available through D041 are low-failure-

---

[2]For this model, we used actual Air Force wartime planning factors (WMP-5).

[3]For a discussion of wartime demand for aircraft parts and deceleration factors see Slay and Sherbrooke (1997).

## Table A.2

### Regression of Number of Removals Against PTOIMDR

| Regression Statistics | |
|---|---|
| Multiple R | 0.45 |
| R square | 0.21 |
| Adjusted R square | 0.19 |
| Standard error | 0.17 |
| Observations | 73 |

| ANOVA | | | | | |
|---|---|---|---|---|---|
| | df | SS | MS | F | Significance F |
| Regression | 1 | 0.511 | 0.511 | 18.659 | 0.000 |
| Residual | 72 | 1.973 | 0.027 | | |
| Total | 73 | 2.484 | | | |

| | Coefficients | Standard error | t stat | P-value |
|---|---|---|---|---|
| Intercept (set to 0) | 0 | | | |
| Number of removals | 0.002254 | 0.000 | 11.465 | 0.000 |

NOTE: SS = sum of squares; MS = mean square.

rate items). Thus, the overall impact of the error in this regression model is quite small.

## TEST STATION CAPACITY

Test station daily supply is the sum of the test station uptime (FMC rate) for each $m^{th}$ tester of $n$ collocated testers multiplied by the hours worked per day.

Test station available time $(x) = \Sigma\ FMC_{nm} *$ hours worked per day

In this analysis, the FMC rates for each tester of $n$ collocated testers are considered to be the same. The previous equation becomes more important if more sophisticated modeling techniques account for the effects of cannibalization.

The overall FMC rate for each tester is estimated through a demand-weighted average of the FMC rates for each LRU tested on the tester.

$$FMC_{nm} = \Sigma\ [FMC_{nmi} * DD_i\ (x)]\ /\ \Sigma\ DD_i\ (x)$$

## Test Station Uptime

Actual uptime by LRU for each test station was available at two bases, Seymour Johnson and Lakenheath. The average uptime of the collocated testers at the bases serves as the average uptime for any number of collocated testers in the model. Depending on tester type, these bases have one, two, or three testers. Therefore, the uptime represents the expected uptime by tester for one, two, or three collocated strings (corresponding to the number of each tester type at these two bases) with the peacetime level of cannibalization practiced by these shops. These data are used to empirically replicate shop behavior rather than trying to fit a theory of cannibalization to the data.

Higher numbers of collocated testers could offer greater cannibalization benefit, so the number of necessary testers for the consolidated cases may be slightly overestimated. If a squadron had to deploy independently with one string, the uptime would likewise probably be lower than used in the model, because there may be some level of cannibalization occurring when all of the squadrons are home at Seymour Johnson or Lakenheath.

## TESTER QUANTITY CALCULATION

The test quantity is set to maximize the capacity utilization subject to a capacity utilization constraint. The maximum capacity utilization factor (CF) depends on the number of test stations and is set to provide a wait time in queue that produces the repair time used in the spare-parts model. Seventy-two hours is the target queue wait time employed in this analysis.

Capacity utilization $(x) = DD(X)$ / test station available time $(x)$

Determining the number of testers:

1. Let $y = 1$

2. Let $n(x) = y$

3. If $CU(x) \leq CFy$, then go to 6

4. Let $y = y + 1$

5. Go to 2

6. $n(x) = y$

## Capacity Utilization Factors

Spare-parts requirements were computed using the Air Force's current ASM and an SBSS-based three-echelon model as appropriate. These models assume an average repair time of about four days to repair an LRU on base. To produce consistency, the capacity utilization factors are designed to produce an average maintenance queue time of 72 hours. For the purposes of the model described here, the capacity factors were set by determining the capacity utilization of an s-server queue with Poisson arrivals and exponential service times (*M/M/s*) that would result in an average queue wait time of 72 hours (see Figure A.3). Standard formulas derived from the birth-and-death process were used. While the actual arrival and service time distributions may be more variable than the Poisson and exponential distributions, this model should still provide close approximations.

$$\rho = \lambda / s\mu < 1 = \text{capacity utilization}$$

$$P_0 = \frac{1}{\left[ \sum_{n=0}^{s-1} \frac{(\lambda/\mu^n)}{n!} + \frac{(\lambda/\mu)^s}{s!} \frac{1}{1-(\lambda/s\mu)} \right]}$$

$$= \text{Prob \{no jobs are waiting for service\}}$$

$$L_q = P_0 (\lambda/\mu) s\rho / (s!(1-\rho)^2) = \text{average length of queue}$$

$$W_q = L_q / \lambda = \text{average wait time in queue}[4]$$

The model uses an assumed mean service time, $\mu$, of 7.41 hours, the demand-weighted average service time estimate across all LRUs. The maximum capacity utilization for a given number of servers is determined by setting $W_q$ to a constant and then solving for the

---

[4]These equations can be found in Hillier and Lieberman (1986) and numerous other sources.

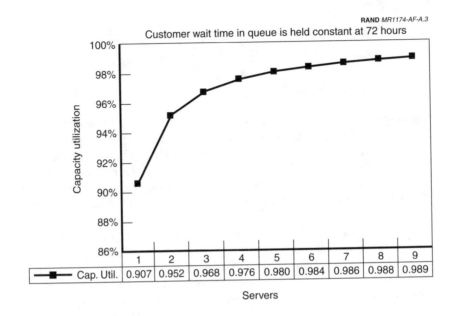

Figure A.3—Capacity Utilization Factors for 72-Hour Wait Time

demand rate that results in this desired $W_q$. The capacity utilization constraint for the given number of servers becomes this demand rate, $\lambda$, divided by the service rate, $s\mu$.

Figure A.4 displays the capacity utilization factors used for the reduced maintenance wait time excursion on page 88 with results in Figure 6.2.

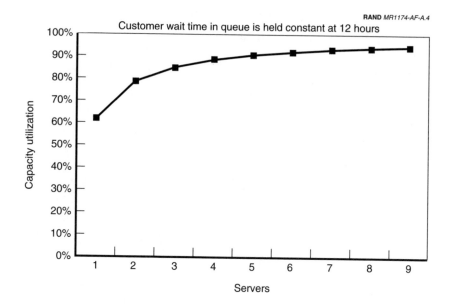

**Figure A.4 — Capacity Utilization Factors for 12-Hour Wait Time**

# TESTER REQUIREMENTS MODEL DATA

The tester requirements model requires a multitude of data elements describing test station demand rates, demand times, and reliability. This appendix provides the data set used to generate the results in this report and describes the sources of the data.

## MODEL DATA

BCS times are expressed in hours. The F-15 SPO column is the BCS time reported by the F-15 SPO (developed through base data collection). The base average column is the average of the BCS times reported by Lakenheath, Seymour Johnson, and Spangdahlem in another data collection effort.

Lakenheath and Seymour Johnson totals are the sum of counts at Lakenheath and Seymour Johnson over six months.

## TESTER RELIABILITY DATA SOURCES

- **Lakenheath:** Six months of test station uptime by LRU by day for EARTS, TISS, EAU, AIS (microwave, displays, computers, I&C, CNI), and METS

- **Seymour Johnson:** Five months of test station uptime by LRU for EARTS/ANT A/B, TISS, AIS (microwave, displays), and METS

- Number of collocated testers used to produce model uptime estimates:

EARTS/ANT: 2
TISS: 3
EAU: 3
Microwave: 3
Displays: 3
METS: 2
Computers: 1
CNI: 1
I&C: 1

# Table B.1
## Model Data

| | Identification | | | BCS Time | | Demand | | Lakenheath and Seymour Johnson Totals | | | | Level Maint. Work | |
| --- | --- | --- | --- | --- | --- | --- | --- | --- | --- | --- | --- | --- | --- |
| Test Station | LRU Nomenclature | Master NSN | WUC | Model (Max of SPO and Bases) | F-15 SPO | Base Avg. | DO41/Est. PTOIM DR | Re-movals | BCS | Base Repairs | AWP | Level | Uptime |
| EARTS | 74GU0 - 031 Rdr Ant | 5985-01-277-8913 | 74FU0 | 7.00 | 6.50 | 7.00 | 0.580 | 304 | 5 | 217 | 92 | 3 | 92% |
| EARTS | 74GH0 - 610 LVPS | 6130-01-123-4126 | 74FH0 | 3.20 | 3.20 | 2.65 | 0.414 | 106 | 18 | 49 | 38 | 3 | 97% |
| EARTS | 74GA0 - 111 Rdr Xmtr | 5841-01-100-8055 | 74GA0 | 5.00 | 5.00 | 3.30 | 0.467 | 207 | 37 | 142 | 32 | 3 | 87% |
| EARTS | 011 | 5841-01-100-7363 | 74FA0 | 5.00 | 5.00 | | 0.689 | | | | | 3 | 87% |
| TISS | 76BA0 - LRU-2C | 6130-01-290-5835 | 76BA0 | 1.83 | 0.30 | 1.83 | 0.232 | 100 | 33 | 47 | 14 | 3 | 94% |
| TISS | 76BB0 - LRU-3C | 5865-01-287-6182 | 76BB0 | 4.73 | 1.10 | 4.73 | 0.305 | 140 | 15 | 97 | 26 | 3 | 91% |
| TISS | 76BD0 - LRU-6C | 5865-01-179-9699 | 76BD0 | 6.08 | 0.70 | 6.08 | 0.232 | 99 | 20 | 63 | 12 | 3 | 93% |
| TISS | 76GF0 - LRU-201 | 5865-01-211-2335 | 76GF0 | 5.60 | 1.80 | 5.60 | 0.205 | 69 | 3 | 39 | 14 | 3 | 90% |
| TISS | B1 CO | 5865-01-100-3768 | 76HB0 | 2.75 | 1.10 | 2.75 | 0.287 | 11 | 2 | 4 | 2 | 3 | 99% |
| TISS | B1 RFA | 5996-01-045-6276 | 76HA0 | 2.00 | 0.45 | 2.00 | 0.444 | 9 | 3 | 3 | 1 | 3 | 99% |
| TISS | 76HG0 - B2 CO | 5865-01-242-8918 | 76HG0 | 2.00 | 1.20 | 2.00 | 0.196 | 18 | 4 | 13 | 5 | 3 | 99% |
| TISS | B2 RFA | 5996-01-242-8347 | 76HF0 | 1.75 | 0.35 | 1.75 | 0.190 | 4 | 2 | 2 | 1 | 3 | 99% |
| TISS | 76MF0 - B3 CO | 5895-01-413-9798 | 76MF0 | 6.20 | 1.70 | 6.20 | 0.444 | 108 | 23 | 71 | 14 | 3 | 92% |
| TISS | 76MA0 - B3 RFA | 5996-01-345-1134 | 76MA0 | 4.73 | 0.75 | 4.73 | 0.248 | 128 | 15 | 75 | 26 | 3 | 92% |
| TISS | 76MFU - B3 RTU | 5895-01-338-8850 | 76MFU | 0.75 | 0.75 | | 0.042 | 13 | 6 | 4 | 2 | 3 | 99% |
| TISS | B1 B2 TU | 5865-00-605-6182 | 76HGD | 0.45 | 0.45 | | 0.011 | 5 | 4 | 1 | 1 | 3 | 100% |
| TISS | LRU 4 PROM | 5865-01-034-5861 | 76HGA | 0.50 | 0.50 | | 0.000 | | | | | 3 | 96% |
| TISS | LRU-9 | 5865-01-289-9651 | 76AG0 | 1.75 | 0.35 | 1.75 | 0.011 | 5 | 2 | 3 | 2 | 3 | 100% |
| EAU | DEEC | 2995-01-339-2207 | 23HF0 | 2.00 | 2.00 | 1.45 | 0.149 | 66 | 41 | 17 | 8 | 3 | 100% |
| EAU | EDU | 6620-01-371-5040 | 23KC0 | 2.00 | 2.00 | 1.50 | 0.291 | 129 | 39 | 76 | 17 | 3 | 100% |
| Displays | 74KF0 - WFOV HUD | 1270-01-232-9337 | 74KF0 | 2.90 | 2.00 | 2.90 | 0.361 | 166 | 20 | 107 | 21 | 3 | 63% |
| Displays | 74KA0 - HUD | 1270-01-183-8987 | 74KA0 | 6.00 | 6.00 | 4.00 | 0.547 | 0 | 0 | 0 | 0 | 3 | 60% |
| Displays | 74KC0 - HUD SDP | 1270-01-040-5948 | 74KC0 | 3.00 | 3.00 | 1.50 | 0.225 | 0 | 0 | 0 | 0 | 3 | 77% |
| Displays | 74MC0 - MPD | 1270-01-230-8578 | 74MC0 | 2.15 | 2.00 | 2.15 | 0.155 | 300 | 56 | 182 | 27 | 3 | 94% |
| Displays | 74MA/B0 - MPCD | 5895-01-224-7827 | 74MB0 | 4.00 | 4.00 | 2.23 | 0.125 | 141 | 17 | 88 | 25 | 3 | 92% |
| Displays | 79MG0 - MPDP | 1270-01-384-1108 | 74MG0 | 3.20 | 2.00 | 3.20 | 0.569 | 254 | 105 | 114 | 15 | 3 | 94% |
| Displays | 74JA0 - ANMI/VSD | 6610-01-084-8224 | 74JA0 | 4.00 | 4.00 | 2.00 | 0.373 | 0 | 0 | 0 | 0 | 3 | 66% |

## Table B.1 (continued)

| Identification | | | | BCS Time | | Demand | | Lakenheath and Seymour Johnson Totals | | | | | |
| --- | --- | --- | --- | --- | --- | --- | --- | --- | --- | --- | --- | --- | --- |
| Test Station | LRU Nomenclature | Master NSN | WUC | Model (Max of SPO and Bases) | F-15 SPO | Base Avg. | DO41/ Est. PTOIM DR | Re-movals | BCS | Base Repairs | AWP | Level Maint. Work Level | Uptime |
| Displays | 74JF0 - PSDP | 6610-01-154-5493 | 74JF0 | 3.00 | 3.00 | 2.50 | 0.435 |  | 0 | 0 | 0 | 3 | 68% |
| Displays | PAC CP | 1270-01-236-8438 | 75PA0 | 4.60 | 3.00 | 4.60 | 0.278 | 103 | 19 | 67 | 13 | 3 | 76% |
| Displays | 57AC0 - MSIP CC | 1270-01-327-7834 | 57AC0 | 1.00 |  | 1.00 | 0.375 | 0 | 0 | 0 | 0 | 3 | 90% |
| Displays | VHSIC CC | 1270-01-422-5778 | 57AE0 | 1.00 | 0.50 | 1.00 | 0.186 | 36 | 1 | 16 | 0 | 2 | 95% |
| Displays | 79GQ0 - 082 | 5841-01-376-0002 | 74GQ0 | 2.00 | 2.00 |  | 0.243 | 13 | 6 | 3 | 2 | 3 | 91% |
| Displays | 042 | 5841-01-158-2818 | 74FY0 | 5.00 | 5.00 |  | 0.270 |  |  |  |  | 3 | 91% |
| Displays | 74GF0 - 044 | 5841-01-278-5146 | 74GF0 | 1.53 | 1.50 | 1.53 | 0.264 | 125 | 56 | 50 | 14 | 3 | 91% |
| Displays | 76KC0 - DSA | 5865-01-172-9448 | 76KC0 | 2.75 | 1.00 | 2.75 | 0.023 | 65 | 3 | 16 | 0 | 2 | 95% |
| Displays | ESCP/CMD | 5865-01-173-6012 | 76KA0 | 1.50 | 1.00 | 1.50 | 0.047 | 29 | 6 | 13 | 0 | 2 | 95% |
| Displays | 65BH0 - IRE | 5895-01-016-2209 | 65BH0 | 4.00 | 4.00 | 2.75 | 0.244 | 30 | 5 | 17 | 9 | 3 | 81% |
| Displays | 44BFA - MRA | 1680-01-015-1783 | 44BFA | 2.50 | 2.50 | 1.13 | 0.005 | 2 | 0 | 1 | 2 | 3 | 94% |
| Displays | 13HA0 - SKID | 1630-01-018-2004 | 13HA0 | 2.00 | 2.00 | 1.18 | 0.095 | 24 | 13 | 1 | 0 | 2 | 84% |
| Displays | FDA | 1680-01-065-2355 | 49AAC | 1.50 | 1.50 | 1.00 | 0.020 | 0 | 0 | 0 | 0 | 2 | 77% |
| Displays | 51NA0 - HSI | 6605-01-042-4831 | 51NA0 | 1.00 | 1.00 | 0.75 | 0.425 | 0 | 0 | 0 | 0 | 2 | 77% |
| Microwave | 74GC0 - 025 | 5841-01-315-0646 | 74GC0 | 4.55 | 3.00 | 4.55 | 0.343 | 156 | 43 | 90 | 19 | 2 | 46% |
| Microwave | 74GS0 - 038 | 5895-01-297-3689 | 74GS0 | 3.75 | 2.00 | 3.75 | 0.382 | 201 | 65 | 80 | 32 | 3 | 54% |
| Microwave | 74GQ0 - 083 | 5841-01-373-3437 | 74GQ0 | 2.00 | 2.00 | 1.70 | 0.243 | 93 | 42 | 34 | 8 | 3 | 81% |
| Microwave | 76CA0 - IB | 5895-01-240-4455 | 76CA0 | 1.53 | 1.00 | 1.53 | 0.020 | 4 | 0 | 0 | 0 | 2 | 71% |
| Microwave | 76KC0 - DSA | 5865-01-172-9448 | 76KC0 | 1.00 | 1.00 | 1.00 | 0.023 | 24 | 3 | 5 | 0 | 2 | 71% |
| Microwave | ESCP/CMD | 5865-01-173-6012 | 76KA0 | 1.00 | 1.00 |  | 0.047 | 13 | 3 | 4 | 0 | 2 | 76% |
| Microwave | 74MC0 - MPD | 1270-01-230-8578 | 74MC0 | 4.00 | 4.00 | 1.00 | 0.155 | 51 | 13 | 26 | 5 | 3 | 82% |
| Microwave | 74MA/B0 - MPCD | 5895-01-227-8102 | 74MA0 | 4.00 | 4.00 | 1.25 | 0.219 | 0 | 0 | 0 | 0 | 3 | 82% |
| Microwave | 74MG0 - MPDP | 1270-01-384-1108 | 74MG0 | 2.00 | 2.00 | 2.00 | 0.569 | 37 | 26 | 8 | 1 | 3 | 81% |
| Microwave | 57AC0 - MSIP CC | 1270-01-327-7834 | 57AC0 | 1.00 |  | 1.00 | 0.375 | 0 | 0 | 0 | 0 | 3 | 81% |
| Microwave | RFO | 5955-01-003-2850 | 74FI0 | 3.00 | 3.00 |  | 0.325 |  |  |  |  | 3 | 82% |
| Microwave | 039 | 5841-01-135-6194 | 74FS0 | 3.00 | 3.00 |  | 0.573 |  |  |  |  | 3 | 82% |
| Microwave | 081 | 5841-01-234-8535 | 74FQ0 | 2.50 | 2.50 |  | 0.601 |  |  |  |  | 3 | 82% |
| Microwave | 022 | 5841-01-048-6312 | 74FC0 | 3.00 | 3.00 |  | 0.463 |  |  |  |  | 3 | 82% |
| Microwave | VHSIC CC | 1270-01-422-5778 | 57AE0 | 0.50 | 0.50 |  | 0.186 | 0 | 0 | 0 | 0 | 2 | 81% |

**Table B.1 (continued)**

| | Identification | | | BCS Time | | Demand | | Lakenheath and Seymour Johnson Totals | | | | Level Maint. Work Level | Uptime |
|---|---|---|---|---|---|---|---|---|---|---|---|---|---|
| Test Station | LRU Nomenclature | Master NSN | WUC | Model (Max of SPO and Bases) | F-15 SPO | Base Avg. | DO41/ Est. PTOIM DR | Re-movals | BCS | Base Repairs | AWP | | |
| Computers | 76KC0 - DSA | 5865-01-172-9448 | 76KC0 | 1.00 | 1.00 | 1.00 | 0.023 | 24 | 3 | 5 | 0 | 2 | 95% |
| Computers | ESCP | 5865-01-173-6012 | 76KA0 | 1.00 | 1.00 | | 0.047 | 13 | 3 | 4 | 0 | 2 | 95% |
| Computers | 74MC0 - MPD | 1270-01-230-8578 | 74MC0 | 1.50 | 1.50 | 1.00 | 0.155 | 51 | 13 | 26 | 5 | 3 | 95% |
| Computers | 74MA/B0 - MPCD | 5895-01-227-8102 | 74MA0 | 4.00 | 4.00 | 1.25 | 0.219 | 0 | 0 | 0 | 0 | 3 | 95% |
| Computers | 79MG0 - MPDP | 1270-01-384-1108 | 74MG0 | 2.00 | 2.00 | | 0.569 | 37 | 26 | 8 | 1 | 3 | 94% |
| Computers | 52AL0 - ASA | 6615-00-262-4314 | 52AL0 | 1.00 | 1.00 | 0.50 | 0.108 | 0 | 0 | 0 | 0 | 2 | 91% |
| Computers | ACCEL IND | 6610-00-361-6686 | 51AM0 | 1.00 | 1.00 | | 0.019 | 0 | 0 | 0 | 0 | 2 | 95% |
| Computers | 51AD0 - AI | 6610-01-167-6617 | 51AD0 | 1.00 | 1.00 | 0.75 | 0.120 | 0 | 0 | 0 | 0 | 2 | 73% |
| Computers | LEA | 6610-00-525-3305 | 55CB0 | 1.00 | 1.00 | | 0.049 | 14 | 7 | 0 | 0 | 2 | 95% |
| Computers | 51AJ0 - STBY AI | 6610-00-160-0905 | 51AJ0 | 1.00 | 1.00 | 1.00 | 0.035 | 4 | 0 | 0 | 0 | 2 | 95% |
| Computers | STBY AIR SPD | 6610-00-296-3574 | 51AG0 | 1.00 | 1.00 | | 0.057 | 5 | 0 | 0 | 0 | 2 | 95% |
| Computers | 51AH0 - PRESS ALT | 6610-00-329-3495 | 51AH0 | 1.00 | 1.00 | 1.00 | 0.061 | 3 | 0 | 0 | 0 | 2 | 91% |
| Computers | COMP ALT | 6685-00-333-6763 | 41AAU | 1.00 | 1.00 | | 0.013 | 0 | 0 | 0 | 0 | 2 | 95% |
| Computers | 52AC0 - RSA | 6615-00-137-7514 | 52AC0 | 2.00 | 2.00 | | 0.226 | 0 | 0 | 0 | 0 | 2 | 70% |
| Computers | 41ABL - ACAC | 1660-01-137-4105 | 41ABL | 2.50 | 2.50 | 0.50 | 0.087 | 0 | 0 | 0 | 0 | 2 | 73% |
| Computers | CCAC | 1660-01-080-8229 | 41AAC | 1.00 | 1.00 | | 0.098 | 0 | 0 | 0 | 0 | 2 | 73% |
| Computers | 51EA0 - ADC | 6610-01-037-9144 | 51EA0 | 3.00 | 3.00 | 2.25 | 0.182 | 41 | 16 | 5 | 1 | 2 | 70% |
| Computers | 57AC0 - MSIP CC | 1270-01-327-7834 | 57AC0 | 1.00 | | 1.00 | 3.750 | 0 | 0 | 5 | 0 | 3 | 95% |
| Computers | VHSIC CC | 1270-01-422-5778 | 57AE0 | 0.50 | 0.50 | | 0.186 | 0 | 0 | 0 | 0 | 2 | 95% |
| Computers | 55CA0 - CDU | 6610-01-018-2431 | 55CA0 | 2.50 | 2.50 | 2.00 | 0.044 | 5 | 3 | 0 | 0 | 2 | 84% |
| Computers | 71FB0 - DG | 6615-00-303-6728 | 71FB0 | 2.00 | 1.50 | 2.00 | 0.250 | 12 | 7 | 0 | 0 | 2 | 95% |
| Computers | 52AM0 - DPS | 6615-01-084-4995 | 52AM0 | 1.00 | 1.00 | | 0.000 | 0 | 0 | 0 | 0 | 2 | 92% |
| Computers | 51EF0 - EAIC | 6610-01-342-9774 | 51EF0 | 2.50 | 2.50 | 2.00 | 0.085 | 18 | 6 | 5 | 1 | 2 | 70% |
| Computers | 49CAA - EID | 6340-00-332-7300 | 49CAA | 1.50 | 1.50 | 1.00 | 0.071 | 2 | 1 | 0 | 0 | 2 | 73% |
| Computers | 52AA0 - PITCH | 6615-01-015-4794 | 52AA0 | 5.00 | 5.00 | | 0.234 | 0 | 0 | 0 | 0 | 2 | 91% |
| Computers | 52AB0 - ROLL/YAW | 6615-01-148-4182 | 52AB0 | 6.00 | 6.00 | | 0.215 | 0 | 0 | 0 | 0 | 2 | 91% |
| Computers | 74EB0 - LCG | 1270-01-046-9884 | 74EB0 | 2.50 | 2.50 | 2.00 | 0.313 | 0 | 0 | 0 | 0 | 2 | 73% |
| Computers | 71FE0 - MAD | 6605-00-314-2536 | 71FE0 | 1.00 | 1.00 | 1.00 | 0.058 | 2 | 0 | 0 | 0 | 2 | 95% |
| Computers | 71AK0 - NCI | 6605-01-094-0775 | 71AK0 | 2.50 | 2.50 | 1.00 | 1.147 | 0 | 0 | 0 | 0 | 2 | 93% |

Table B.1 (continued)

| Test Station | LRU Nomenclature | Master NSN | WUC | BCS Time Model (Max of SPO and Bases) | BCS Time F-15 SPO | Demand Base Avg. | Demand DO41/Est. PTOIM DR | Re-movals | BCS | Base Repairs | AWP | Level Maint. Work Level | Uptime |
|---|---|---|---|---|---|---|---|---|---|---|---|---|---|
| Computers | 55BC0 - SDR | 6610-00-349-4184 | 55BC0 | 3.00 | 3.00 | 3.00 | 0.522 | 16 | 2 | 2 | 0 | 2 | 69% |
| Computers | 71FA0 - ECA | 6605-00-149-1134 | 71FA0 | 2.50 | 2.50 | 2.25 | 0.343 | 18 | 10 | 0 | 0 | 2 | 85% |
| METS | 52BA0 - FCC | 6615-01-411-0566 | 52BA0 | 2.45 | 2.00 | 2.45 | 0.426 | 189 | 101 | 70 | 10 | 3 | 86% |
| METS | 57CA0 - AIU #2 | 1270-01-231-6341 | 57CA0 | 1.00 | 1.00 | 1.00 | 0.170 | 11 | 5 | 5 | 2 | 2 | 100% |
| METS | 57CB0 - AIU #1 | 1270-01-356-2585 | 57CB0 | 1.35 | 1.00 | 1.35 | 0.070 | 86 | 25 | 51 | 8 | 3 | 100% |
| METS | 57CB0 - AIU #2 | 1270-01-231-4064 | 57CB0 | 1.30 | 1.00 | 1.30 | 0.070 | 15 | 8 | 9 | 1 | 3 | 100% |
| METS | 82AA0 - RMR | 6605-01-240-0136 | 82AA0 | 3.48 | 1.50 | 3.48 | 0.701 | 344 | 35 | 269 | 23 | 3 | 82% |
| METS | 57DA0 - UFCP | 5895-01-306-2073 | 57DA0 | 1.15 | 0.50 | 1.15 | 0.235 | 213 | 20 | 151 | 52 | 3 | 100% |
| METS | 51BA0 - EMD | 6620-01-232-0680 | 51BA0 | 0.70 | 0.50 | 0.70 | 0.229 | 105 | 37 | 47 | 15 | 3 | 100% |
| METS | 63BU0 - ICSCP | 5895-01-382-3225 | 63BU0 | 1.50 | 1.00 | 1.50 | 0.109 | 34 | 12 | 13 | 4 | 3 | 86% |
| METS | 63AV0 - UHF R/T | 5821-01-228-7058 | 63AV0 | 1.63 | 1.00 | 1.63 | 0.164 | 143 | 25 | 51 | 0 | 2 | 95% |
| METS | 65AA0 - IFF R/T | 5895-01-112-6380 | 65AA0 | 6.40 | 6.40 | 2.30 | 0.176 | 49 | 7 | 7 | 0 | 2 | 90% |
| METS | 65BC/D0 - AAI R/T | 5895-01-273-1990 | 65BD0 | 2.50 | 1.00 | 2.50 | 0.423 | 109 | 6 | 37 | 0 | 2 | 94% |
| METS | 71FA0 - ECA | 6605-00-149-1134 | 71FA0 | 2.50 | 2.50 | 2.07 | 0.343 | 43 | 11 | 7 | 0 | 2 | 100% |
| METS | ADF ECA | 5996-00-262-5018 | 71BD0 | 4.00 | 4.00 | 0.50 | 0.006 | 3 | 0 | 1 | 1 | 2 | 100% |
| METS | 71CA0 - ILS RCVR | 5826-01-021-1744 | 71CA0 | 4.00 | 4.00 | 1.93 | 0.064 | 10 | 0 | 1 | 0 | 2 | 93% |
| METS | 71ZA0 - TACAN R/T | 5826-01-012-1938 | 71ZA0 | 2.43 | 1.50 | 2.43 | 0.107 | 17 | 0 | 1 | 0 | 2 | 85% |
| METS | 71ZF0 - TACAN MNT | 5826-01-060-3893 | 71ZF0 | 1.70 | 1.00 | 1.70 | 0.059 | 5 | 2 | 0 | 0 | 2 | 85% |
| METS | 231AM - NOZ POS | 6620-01-195-9950 | 231AM | 1.50 | 1.50 | 1.25 | 0.053 | 23 | 4 | 0 | 0 | 2 | 100% |
| METS | 74EB0 - LCG | 1270-01-046-9884 | 74EB0 | 2.50 | 2.50 | 2.00 | 0.313 | 0 | 0 | 0 | 0 | 2 | 93% |
| METS | 51EF0 - EAIC | 6610-01-342-9774 | 51EF0 | 2.50 | 2.50 | 2.00 | 0.085 | 102 | 32 | 45 | 13 | 2 | 46% |
| METS | 71FB0 - DG | 6615-00-303-6728 | 71FB0 | 1.67 | 1.50 | 1.67 | 0.250 | 33 | 11 | 0 | 0 | 2 | 93% |
| METS | 51EA0 - ADC | 6610-01-037-9144 | 51EA0 | 3.00 | 3.00 | 2.13 | 0.182 | 41 | 16 | 5 | 1 | 2 | 97% |
| METS | LEA | 6610-00-525-3305 | 55CB0 | 1.00 | 1.00 | | 0.049 | 0 | 0 | 0 | 0 | 2 | 93% |
| Indicators & Controls | 63BE0 - AAI C/P | 5895-01-044-4987 | 63BE0 | 3.00 | 3.00 | 0.50 | 0.037 | 0 | 0 | 0 | 0 | 2 | 93% |
| Indicators & Controls | 231AP - OPT | 6620-00-290-6505 | 231AP | 2.00 | 2.00 | 0.88 | 0.002 | 1 | 1 | 0 | 0 | 2 | 83% |

**Table B.1 (continued)**

| Test Station | LRU Nomenclature | Master NSN | WUC | Model (Max of SPO and Bases) | F-15 SPO | Base Avg. | DO41/ Est. PTOIM DR | Re-movals | BCS | Base Repairs | AWP | Level Maint. Work Level | Uptime |
|---|---|---|---|---|---|---|---|---|---|---|---|---|---|
| | | | | BCS Time | | Demand | | Lakenheath and Seymour Johnson Totals | | | | | |
| Indicators & Controls | 45CDE - HPT | 6685-01-433-3057 | 45CDE | 0.75 | | 0.75 | 0.026 | 18 | 8 | 5 | 0 | 3 | 82% |
| Indicators & Controls | 51AL0 - AOA IND | 6320-00-134-2251 | 51AL0 | 2.00 | 2.00 | 0.25 | 0.000 | 0 | 0 | 0 | 0 | 2 | 91% |
| Indicators & Controls | 51ED0 - AOA XMTR | 6610-00-535-7722 | 51ED0 | 2.00 | 2.00 | 1.00 | 0.077 | 6 | 0 | 0 | 0 | 2 | 80% |
| Indicators & Controls | 55AB0 - ASP | 1680-01-162-5850 | 55AB0 | 3.00 | 3.00 | 1.00 | 0.031 | 2 | 1 | 0 | 0 | 2 | 92% |
| Indicators & Controls | 55AE0 - BIT C/P | 1680-01-157-2424 | 55AE0 | 3.50 | 3.50 | 0.75 | 0.149 | 0 | 0 | 0 | 0 | 2 | 80% |
| Indicators & Controls | 71FC0 - CMPS C/P | 6615-00-303-6730 | 71FC0 | 1.00 | 1.00 | 1.00 | 0.049 | 0 | 0 | 0 | 0 | 2 | 93% |
| Indicators & Controls | ENG CNTL | 6615-01-021-4234 | 52AH0 | 2.00 | 2.00 | | 0.049 | 0 | 0 | 0 | 0 | 2 | 80% |
| Indicators & Controls | ELEC TACH | 6620-00-148-7306 | 231AA | 2.00 | 2.00 | | 0.129 | 0 | 0 | 0 | 0 | 2 | 80% |
| Indicators & Controls | 46ECD - FLCU | 6680-00-567-8668 | 46EC0 | 1.00 | 1.00 | | 0.000 | 0 | 0 | 0 | 0 | 3 | 81% |
| Indicators & Controls | FLTC | 2915-01-061-3522 | 0 | | | | 0.000 | 0 | 0 | 0 | 0 | 0 | 93% |
| Indicators & Controls | 231AB - FTIT | 6685-01-048-2889 | 231AB | 1.00 | 1.00 | 0.50 | 0.100 | 0 | 0 | 0 | 0 | 2 | 93% |
| Indicators & Controls | 46EDA - FFI | 6620-00-468-9824 | 46EDA | 1.00 | 1.00 | 0.50 | 0.038 | 0 | 0 | 0 | 0 | 2 | 91% |
| Indicators & Controls | 46EBA - FQI | 6680-01-103-3419 | 46EBA | 3.50 | 3.50 | 0.88 | 0.092 | 14 | 6 | 0 | 0 | 2 | 93% |
| Indicators & Controls | 46EBB - SIG COND | 6680-01-106-6215 | 46EBB | 3.50 | 3.50 | 0.75 | 0.092 | 11 | 0 | 1 | 0 | 2 | 46% |

**Table B.1 (continued)**

| Test Station | Identification | | | BCS Time | | Demand | | Lakenheath and Seymour Johnson Totals | | | | Level Maint. Work Level | Uptime |
| | LRU Nomenclature | Master NSN | WUC | Model (Max of SPO and Bases) | F-15 SPO | Base Avg. | DO41/ Est. PTOIM DR | Re-movals | BCS | Base Repairs | AWP | | |
| --- | --- | --- | --- | --- | --- | --- | --- | --- | --- | --- | --- | --- | --- |
| Indicators & Controls | 63BJ0 - ICCP | 5895-01-329-6324 | 63BJ0 | 5.50 | 5.50 | 3.00 | 0.676 | 0 | 0 | 0 | 0 | 2 | 79% |
| Indicators & Controls | 63BF0 - IFF C/P | 5895-00-340-9619 | 63BF0 | 3.00 | 3.00 | 0.50 | 0.040 | 0 | 0 | 0 | 0 | 2 | 79% |
| Indicators & Controls | 76AF0 - TEWS C/P | 5865-00-477-5704 | 76AF0 | 0.25 | | 0.25 | 0.035 | 5 | 1 | 3 | 2 | 3 | 93% |
| Indicators & Controls | CAMERA | 6710-01-018-2007 | 74KE0 | 2.80 | 2.80 | | 0.000 | | | | | 2 | 84% |
| Indicators & Controls | RDR SET CONT (MSIP/70) | 5841-01-193-8113 | 74GK0 | 2.10 | 2.10 | | 0.000 | | | | | 2 | 84% |
| Indicators & Controls | RDR SET CONT (APG63) | 5841-01-058-8862 | 74FK0 | 2.10 | 2.10 | | 0.053 | | | | | 2 | 84% |
| Indicators & Controls | TCCP | 5895-00-349-6061 | 63CB0 | 3.00 | 3.00 | | 0.083 | | | | | 2 | 84% |
| Indicators & Controls | MCCP | 5895-01-095-9593 | 63BH0 | 3.00 | 3.00 | | 0.212 | | | | | 2 | 84% |
| Indicators & Controls | ALTITUDE INDICATOR | 6610-00-000-0122 | 51AK0 | 2.50 | 2.50 | | 0.101 | | | | | 2 | 84% |
| Indicators & Controls | VERTICAL SPEED INDICATOR | 6610-00-134-2259 | 51AF0 | 1.00 | 1.00 | | 0.046 | | | | | 2 | 84% |
| Indicators & Controls | AIRSPEED MACH | 6610-00-134-2260 | 51AE0 | 2.00 | 2.00 | | 0.035 | | | | | 2 | 84% |
| Indicators & Controls | LEVEL SENS CONTROL | 2915-00-567-8668 | 46ECD | 1.00 | 1.00 | | 0.000 | | | | | 2 | 84% |
| Indicators & Controls | INT LIGHTS P/S | 1680-01-062-6828 | 44BF0 | 2.50 | 2.50 | | 0.000 | | | | | 2 | 84% |
| Indicators & Controls | CLDU | 1680-01-048-5183 | 44EA0 | 1.00 | 1.00 | | 0.006 | | | | | 2 | 84% |

Table B.1 (continued)

| | Identification | | | BCS Time | | Demand | | Lakenheath and Seymour Johnson Totals | | | | | |
|---|---|---|---|---|---|---|---|---|---|---|---|---|---|
| Test Station | LRU Nomenclature | Master NSN | WUC | Model (Max of SPO and Bases) | F-15 SPO | Base Avg. | DO41/ Est. PTOIM DR | Re-movals | BCS | Base Repairs | AWP | Level Maint. Work Level | Uptime |
| Indicators & Controls | CLLU | 1680-01-032-5251 | 44EC0 | 3.50 | 3.50 | | 0.049 | | | | | 2 | 84% |
| Indicators & Controls | GCU | 6110-01-049-8639 | 42AF0 | 8.00 | 8.00 | | 0.000 | | | | | 2 | 84% |
| Indicators & Controls | MGCU | 6610-00-539-0411 | 42FF0 | 2.50 | 2.50 | | 0.025 | | | | | 2 | 84% |
| Indicators & Controls | RH THROTTLE GRIP | 1680-01-159-6742 | 231FN | 1.00 | 1.00 | | 0.365 | | | | | 2 | 84% |
| Indicators & Controls | NOZ POS XMTR | 6620-00-124-9515 | 231AM | 1.50 | 1.50 | | 0.000 | | | | | 2 | 84% |
| Indicators & Controls | NOZ POS IND | 6620-00-312-2036 | 231AC | 1.50 | 1.50 | | 0.015 | | | | | 2 | 84% |
| Indicators & Controls | OIL PRESSURE IND | 6620-00-531-6100 | 231AG | 1.00 | 1.00 | | 0.000 | | | | | 2 | 84% |
| Indicators & Controls | PITCH RATIO IND | 6610-00-303-6706 | 14ADF | 1.00 | 1.00 | | 0.011 | | | | | 2 | 84% |
| Indicators & Controls | LANDING GEAR CONTROL | 1680-01-135-5626 | 13FA0 | 2.00 | 2.00 | | 0.000 | | | | | 2 | 84% |
| Indicators & Controls | 231FJ - LHTG | 1680-01-159-5332 | 231FT | 1.00 | 1.00 | 0.50 | 0.060 | 1 | 0 | 0 | 0 | 2 | 83% |
| Comm, Navigation & Identification | 65BC/D0 - AAI R/T | 5895-01-273-1990 | 65BD0 | 2.00 | 1.00 | 2.00 | 0.423 | 11 | 1 | 4 | 0 | 2 | 98% |
| Comm, Navigation & Identification | ADF ECA | 5996-00-262-5018 | 71BD0 | 4.00 | 4.00 | | 0.006 | 0 | 0 | 0 | 0 | 2 | 98% |
| Comm, Navigation & Identification | ANT SEL | 5985-00-509-2466 | 0 | | | | 0.000 | 0 | 0 | 0 | 0 | 0 | 100% |

## Table B.1 (continued)

| Test Station | Identification | | | BCS Time | | Demand | | Lakenheath and Seymour Johnson Totals | | | | | |
| | LRU Nomenclature | Master NSN | WUC | Model (Max of SPO and Bases) | F-15 SPO | Base Avg. | DO41/ Est. PTOIM DR | Re-movals | BCS | Base Repairs | AWP | Level Maint. Work Level | Uptime |
|---|---|---|---|---|---|---|---|---|---|---|---|---|---|
| Comm, Navigation & Identification | 71CA0 - ILS RCVR | 5826-01-021-1744 | 71CA0 | 4.00 | 4.00 | 0.75 | 0.064 | 4 | 0 | 0 | 0 | 2 | 98% |
| Comm, Navigation & Identification | ILS T/S | 6625-00-004-5378 | 0 | | | | 0.000 | 0 | 0 | 0 | 0 | 0 | 98% |
| Comm, Navigation & Identification | 71ZF0 - TACAN MNT | 5826-01-060-3893 | 71ZF0 | 1.50 | 1.00 | 1.50 | 0.059 | 2 | 1 | 0 | 0 | 2 | 100% |
| Comm, Navigation & Identification | 71ZA0 - TACAN R/T | 5826-01-012-1938 | 71ZA0 | 2.00 | 1.50 | 2.00 | 0.107 | 7 | 0 | 1 | 0 | 2 | 100% |
| Comm, Navigation & Identification | 63AV0 - R/T 1145B | 5821-01-228-7058 | 63AV0 | 1.00 | 1.00 | 0.75 | 0.164 | 44 | 11 | 19 | 0 | 2 | 100% |
| Comm, Navigation & Identification | 65AA0 - XPNDR | 5895-01-112-6380 | 65AA0 | 6.40 | 6.40 | 1.50 | 0.176 | 10 | 4 | 1 | 0 | 2 | 98% |
| ESTS | 74KF0 - WFOV HUD | 1270-01-232-9337 | | 2.90 | 2.00 | 2.90 | 0.361 | 166 | 20 | 107 | 21 | 3 | 95% |
| ESTS | 74KA0 - HUD | 1270-01-183-8987 | | 6.00 | 6.00 | 4.00 | 0.547 | 0 | 0 | 0 | 0 | 3 | 95% |
| ESTS | 74KC0 - HUD SDP | 1270-01-040-5948 | | 3.00 | 3.00 | 1.50 | 0.225 | 0 | 0 | 0 | 0 | 3 | 95% |
| ESTS | 74MC0 - MPD | 1270-01-230-8578 | | 2.15 | 2.00 | 2.15 | 0.155 | 300 | 56 | 182 | 27 | 3 | 95% |
| ESTS | 74MA/B0 - MPCD | 5895-01-224-7827 | | 4.00 | 4.00 | 2.23 | 0.125 | 141 | 17 | 88 | 25 | 3 | 95% |
| ESTS | 79MG0 - MPDP | 1270-01-384-1108 | | 3.20 | 2.00 | 3.20 | 0.569 | 254 | 105 | 114 | 15 | 3 | 95% |
| ESTS | 74JA0 - ANMI/VSD | 6610-01-084-8224 | | 4.00 | 4.00 | 2.00 | 0.373 | 0 | 0 | 0 | 0 | 3 | 95% |
| ESTS | 74JF0 - PSDP | 6610-01-154-5493 | | 3.00 | 3.00 | 2.50 | 0.435 | 0 | 0 | 0 | 0 | 3 | 95% |
| ESTS | PAC CP | 1270-01-236-8438 | | 4.60 | 3.00 | 4.60 | 0.278 | 103 | 19 | 67 | 13 | 3 | 95% |
| ESTS | VHSIC CC | 1270-01-422-5778 | | 1.00 | 0.50 | 1.00 | 0.186 | 36 | 1 | 16 | 0 | 2 | 95% |
| ESTS | 79GQ0 - 082 | 5841-01-376-0002 | | 2.00 | 2.00 | | 0.243 | 13 | 6 | 3 | 2 | 3 | 95% |

## Table B.1 (continued)

| | Identification | | | BCS Time | | Demand | | Lakenheath and Seymour Johnson Totals | | | | | |
|---|---|---|---|---|---|---|---|---|---|---|---|---|---|
| Test Station | LRU Nomenclature | Master NSN | WUC | Model (Max of SPO and Bases) | F-15 SPO | Base Avg. | DO41/ Est. PTOIM DR | Re-movals | BCS | Base Repairs | AWP | Level Maint. Work Level | Uptime |
| ESTS | 042 | 5841-01-158-2818 | | 5.00 | 5.00 | | 0.270 | 125 | 56 | 50 | 14 | 3 | 95% |
| ESTS | 74GF0 - 044 | 5841-01-278-5146 | | 1.53 | 1.50 | 1.53 | 0.264 | 65 | 3 | 16 | 0 | 3 | 95% |
| ESTS | 76KC0 - DSA | 5865-01-172-9448 | | 2.75 | 1.00 | 2.75 | 0.023 | 29 | 6 | 13 | 0 | 2 | 95% |
| ESTS | ECSP/CMD | 5865-01-173-6012 | | 1.50 | 1.00 | 1.50 | 0.047 | 30 | 5 | 17 | 9 | 2 | 95% |
| ESTS | 65BH0 - IRE | 5895-01-016-2209 | | 4.00 | 4.00 | 2.75 | 0.244 | 0 | 0 | 0 | 0 | 3 | 95% |
| ESTS | FDA | 1680-01-065-2355 | | 1.50 | 1.50 | 1.00 | 0.020 | 156 | 43 | 90 | 0 | 2 | 95% |
| ESTS | 74GC0 - 025 | 5841-01-315-0646 | | 4.55 | 3.00 | 4.55 | 0.343 | 201 | 65 | 80 | 19 | 3 | 95% |
| ESTS | 74GS0 - 038 | 5895-01-297-3689 | | 3.75 | 2.00 | 3.75 | 0.382 | 4 | 0 | 0 | 32 | 3 | 95% |
| ESTS | 76CA0 - IB | 5895-01-240-4455 | | 1.53 | 1.00 | 1.53 | 0.020 | 0 | 0 | 0 | 0 | 2 | 95% |
| ESTS | 74MA/B0 - MPCD | 5895-01-227-8102 | | 4.00 | 4.00 | 1.25 | 0.219 | | | | 0 | 3 | 95% |
| ESTS | RFO | 5955-01-003-2850 | | 3.00 | 3.00 | | 0.325 | | | | | 3 | 95% |
| ESTS | 039 | 5841-01-135-6194 | | 3.00 | 3.00 | | 0.573 | | | | | 3 | 95% |
| ESTS | 081 | 5841-01-234-8535 | | 2.50 | 2.50 | | 0.601 | | | | | 3 | 95% |
| ESTS | 022 | 5841-01-048-6312 | | 3.00 | 3.00 | | 0.463 | | | | | 3 | 95% |
| ESTS | 52AL0 - ASA | 6615-00-262-4314 | | 1.00 | 1.00 | 0.50 | 0.108 | 0 | 0 | 0 | 0 | 3 | 95% |
| ESTS | 52AC0 - RSA | 6615-00-137-7514 | | 2.00 | 2.00 | | 0.226 | 0 | 0 | 0 | 0 | 2 | 95% |
| ESTS | 52AA0 - PITCH | 6615-01-015-4794 | | 5.00 | 5.00 | | 0.234 | 0 | 0 | 0 | 0 | 2 | 95% |
| ESTS | 52AB0 - ROLL/YAW | 6615-01-148-4182 | | 6.00 | 6.00 | | 0.215 | 0 | 0 | 0 | 0 | 2 | 95% |
| ESTS | 71AK0 - NCI | 6605-01-094-0775 | | 2.50 | 2.50 | 1.00 | 1.147 | 0 | 0 | 0 | 0 | 2 | 95% |
| ESTS | 71FA0 - ECA | 6605-00-149-1134 | | 2.50 | 2.50 | 2.25 | 0.343 | 18 | 10 | 0 | 0 | 2 | 95% |
| ESTS | 52BA0 - FCC | 6615-01-411-0566 | | 2.45 | 2.00 | 2.45 | 0.426 | 189 | 101 | 70 | 10 | 3 | 95% |
| ESTS | 57CA0 - AIU #2 | 1270-01-231-6341 | | 1.00 | 1.00 | 1.00 | 0.170 | 11 | 5 | 5 | 2 | 3 | 95% |
| ESTS | 57CB0 - AIU #1 | 1270-01-356-2585 | | 1.35 | 1.00 | 1.35 | 0.070 | 86 | 25 | 51 | 8 | 3 | 95% |
| ESTS | 82AA0 - RMR | 6605-01-240-0136 | | 3.48 | 1.50 | 3.48 | 0.701 | 344 | 35 | 269 | 23 | 3 | 95% |
| ESTS | 57DA0 - UFCP | 5895-01-306-2073 | | 1.15 | 0.50 | 1.15 | 0.235 | 213 | 20 | 151 | 52 | 3 | 95% |
| ESTS | 51BA0 - EMD | 6620-01-232-0680 | | 0.70 | 0.50 | 0.70 | 0.229 | 105 | 37 | 47 | 15 | 3 | 95% |
| ESTS | 63BU0 - ICSCP | 5895-01-382-3225 | | 1.50 | 1.00 | 1.50 | 0.109 | 34 | 12 | 13 | 4 | 3 | 95% |
| ESTS | ADF ECA | 5996-00-262-5018 | | 4.00 | 4.00 | 0.50 | 0.006 | 3 | 0 | 1 | 1 | 2 | 95% |
| ESTS | 71CA0 - ILS RCVR | 5826-01-021-1744 | | 4.00 | 4.00 | 1.93 | 0.064 | 10 | 0 | 1 | 0 | 2 | 95% |

## Table B.1 (cointinued)

| Test Station | Identification | | | BCS Time | | Demand | | Lakenheath and Seymour Johnson Totals | | | | Level Maint. Work | Uptime |
| | LRU Nomenclature | Master NSN | WUC | Model (Max of SPO and Bases) | F-15 SPO | Base Avg. | DO41/ Est. PTOIM DR | Re-movals | BCS | Base Repairs | AWP | Level | |
| --- | --- | --- | --- | --- | --- | --- | --- | --- | --- | --- | --- | --- | --- |
| ESTS | 71ZF0 - TACAN MNT | 5826-01-060-3893 | | 1.70 | 1.00 | 1.70 | 0.059 | 5 | 2 | 0 | 0 | 2 | 95% |
| ESTS | 51EF0 - EAIC | 6610-01-342-9774 | | 2.50 | 2.50 | 2.00 | 0.085 | 102 | 32 | 45 | 13 | 2 | 95% |
| ESTS | 51EA0 - ADC | 6610-01-037-9144 | | 3.00 | 3.00 | 2.13 | 0.182 | 41 | 16 | 5 | 1 | 2 | 95% |
| ESTS | 55AE0 - BIT C/P | 1680-01-157-2424 | | 3.50 | 3.50 | 0.75 | 0.149 | 0 | 0 | 0 | 0 | 2 | 95% |
| ESTS | 63BJ0 - ICCP | 5895-01-329-6324 | | 5.50 | 5.50 | 3.00 | 0.676 | 0 | 0 | 0 | 0 | 2 | 95% |
| ESTS | 63BF0 - IFF C/P | 5895-00-340-9619 | | 3.00 | 3.00 | 0.50 | 0.040 | 0 | 0 | 0 | 0 | 2 | 95% |
| ESTS | RDR SET CONT (MSIP/70) | 5841-01-193-8113 | | 2.10 | 2.10 | | 0.000 | 0 | | | | 2 | 95% |
| ESTS | RDR SET CONT (APG63) | 5841-01-058-8862 | | 2.10 | 2.10 | | 0.053 | | | | | 2 | 95% |
| ESTS | CLLU | 1680-01-032-5251 | | 3.50 | 3.50 | | 0.049 | | | | | 2 | 95% |

# MODEL OUTPUT

## MODEL OUTPUT

To allow for the rapid sizing of a number of different locations under different operating parameters, we created a spreadsheet model to automate the demand and supply calculations. The cells on the left (see Figure C.1) are the input cells. The operating tempo inputs

RAND *MR1174-AF-C.1*

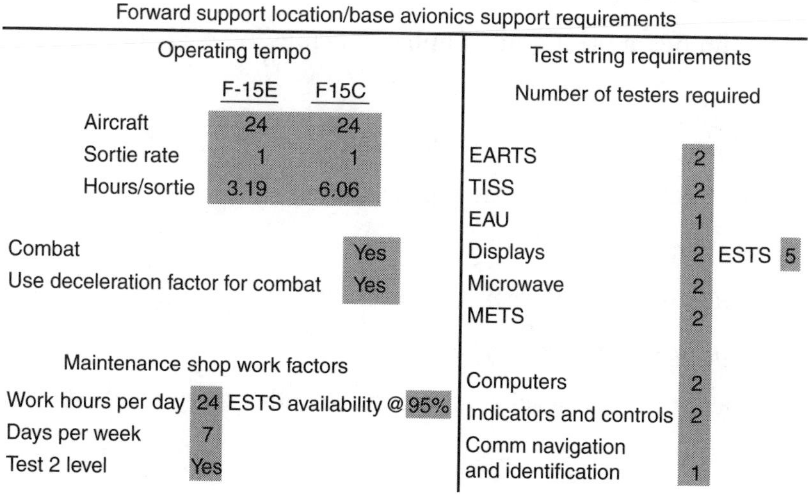

Forward support location/base avionics support requirements

| Operating tempo | | | Test string requirements | |
|---|---|---|---|---|
| | F-15E | F15C | Number of testers required | |
| Aircraft | 24 | 24 | | |
| Sortie rate | 1 | 1 | EARTS | 2 |
| Hours/sortie | 3.19 | 6.06 | TISS | 2 |
| | | | EAU | 1 |
| Combat | | Yes | Displays | 2 | ESTS | 5 |
| Use deceleration factor for combat | | Yes | Microwave | 2 |
| | | | METS | 2 |
| Maintenance shop work factors | | | Computers | 2 |
| Work hours per day 24 ESTS availability @ 95% | | | Indicators and controls | 2 |
| Days per week 7 | | | Comm navigation | |
| Test 2 level Yes | | | and identification | 1 |

Figure C.1—Spreadsheet Model Input and Output Screen

include the number of F-15Es and F-15Cs, the flying hours per day for each based upon sorties per day and hours per sortie, and whether or not they are flying combat or peacetime missions. The combat input cell determines whether the model uses the wartime planning flying hour deceleration factors to compute demand. An additional input cell allows the analyst to model the effect of a combat situation without using the deceleration factors. Shop-work-factor cell inputs include the number of hours worked per day, the number of days worked per week, and whether or not the shop tests two-level items.

The output cells on the right are the number of testers required on the basis of the operating tempo and shop work inputs. Note that the outputs include all three potential tester configurations. For the testers that are substitutes in the different configurations, the quantities for each are alternatives and not additive.

## REQUIRED SPREADSHEET TESTER QUANTITY CALCULATION MODEL INPUTS

Operating tempo:

> Number of F-15Cs and number of F-15Es
>
> Sorties per day (Es and Cs)
>
> Flying hours per sortie (Es and Cs)
>
> Combat status (combat or peacetime)

Supply tempo:

> Days worked per week
>
> Hours worked per day

Whether or not two-level items are tested at the intermediate shop.

"Application of Military Standard Composite Rate Acceleration Factors for Fiscal Year 1998," *AFI 65-503 Cost and Planning Factors*, April 23, 1998, Table A32-1.

Armstrong, Colonel William C., *Operational Requirements Document (ORD) CAF (TAF) 309-87-II/III-B for F-15 Electronic System Test Set Program ACAT Level III* (Draft), LGF, Fighter/Bomber Maintenance Division, ACC, undated.

"Basic Item 217-712, Avionics Shop (ADDM-D)," McDonnel Aircraft Company, MDC, Rev. B, no date.

"F-15 ESTS Beddown Schedule," ESTS Integrated Product Team, ESTS Program Office, Wright-Patterson AFB, Ohio, May 14, 1998.

Galway, Lionel, Robert S. Tripp, Timothy L. Ramey, and John G. Drew, *Supporting Expeditionary Aerospace Forces: New Agile Combat Support Postures*, RAND, MR-1075-AF, 2000.

"Hangar-Aircraft Maintenance," www.buildeval.com/tables/hangar. maint.html, Building Evaluations, Clarksburg, New Jersey, 2000.

Hillier, Frederick S., and Gerald J. Lieberman, *Introduction to Operations Research*, Holden-Day, Inc., Oakland, California, 1986.

Hosek, James, and Mark Totten, *Does Perstempo Hurt Reenlistment? The Effect of Long or Hostile Perstempo on Reenlistment*, RAND, MR-990-OSD, 1998.

O'Malley, T. J., *The Aircraft Availability Model: Conceptual Framework and Mathematics*, Logistics Management Institute, Washington, D.C., June 1983.

Peltz, Eric, Hyman L. Shulman, Robert S. Tripp, and Chief Master Sergeant (S) John G. Drew, "F-15 Avionics Support Options for the Expeditionary Aerospace Force: Technical Appendices," RAND, internal document, January 1999.

"Permanent Change of Station (PCS) Cost Per Move As of FY98/99 President's Budget," *AFI 65-503 Cost and Planning Factors*, March 3, 1998, Table A24-1.

Pyles, Raymond, *The Dyna-METRIC Readiness Assessment Model: Motivation, Capabilities, and Use*, RAND, R-2886-AF, 1984.

Richter, Paul, "Buildup in Gulf Costly: Expenses, Stress Surge for Military," *Los Angeles Times*, November 17, 1998a.

Richter, Paul, "The Tough Job of Keeping Soldiers Ready for War," *Los Angeles Times*, November 22, 1998b.

Ryan, General Michael E., USAF, "Aerospace Expeditionary Force: Better Use of Aerospace Power for the 21st Century," briefing, USAF, Washington, D.C., 1998.

Schnaible, Major Eric, USAF, "EAF Implementation," briefing, HQ USAF/XOPE, Washington, D.C., 1999.

Slay, F. Michael, Tovey C. Bachman, Robert C. Kline, Frank L. Eichorn, and Randall M. King, *Optimizing Spares Support: The Aircraft Sustainability Model*, Logistics Management Institute, AF501MR1, McLean, Virginia, October 1996.

Slay, Michael, and Craig C. Sherbrooke, *Predicting Wartime Demand for Aircraft Spares*, Logistics Management Institute, AF501MR2, McLean, Virginia, April 1997.

Tripp, Robert S., Lionel Galway, Paul S. Killingsworth, Eric Peltz, Timothy L. Ramey, and John Drew, *Supporting Expeditionary Aerospace Forces: An Integrated Strategic Agile Combat Support Planning Framework*, RAND, MR-1056-AF, January 1999.

*USAF Supply Manual*, AFMAN 23-110, Volume II, Part Two, Chapter 19.

Williams, Matthew, "Plea for help (from Air Force secretary and the chief of staff): Better pay, bigger budgets called key to fixing readiness woes," *Air Force Times,* September 28, 1998.